THE BROWN
FOUNDATIONS OF GEOGRAPHY
SERIES

SECOND EDITION

A GEOGRAPHY OF
the atmosphere

JOHN J. HIDORE
Indiana University

WM. C. BROWN COMPANY PUBLISHERS
DUBUQUE, IOWA

Geography is one of man's oldest sciences, yet it is as new as the Space Age. Knowledge of the earth obtained from satellite photography and measurement, remote sensing of the environment, and by means of other sophisticated techniques are really but a stage in the evolutionary process that began with ancient man's curiosity about his surroundings. Man has always been interested in the earth and the things on it. Today this interest may be channeled through the discipline of geography, which offers one means of organizing a vast amount of physical and cultural information.

The **Brown Foundations of Geography Series** has been created to facilitate the study of physical, cultural, and methodological geography at the college level. The **Series** is a carefully selected group of titles that covers the wide spectrum of basic geography. While the individual titles are self-contained, collectively they comprise a modern synthesis of major geographical principles. The underlying theme of each book is to foster an awareness of geography as an imaginative, evolving science.

FOUNDATIONS OF GEOGRAPHY

Consulting Editor
Robert H. Fuson
University of South Florida

Copyright ©1969, 1972 by Wm. C. Brown Company Publishers

Library of Congress Catalog Card Number: 70—178191

ISBN 0—697—05165—X

Printed in the United States of America

Introduction

The environment in which man lives is everchanging. Being dynamic in many respects is due to the irregular flow of energy through the environment. The energy that runs this changing regime comes primarily from three sources: solar radiation, radioactive decomposition, and gravity. The energy in the environment may exist as kinetic energy, potential energy, or latent energy.

The primary or most basic physical system operating in our environment is the relationship between the sun and planet earth. Energy flows from the sun to the earth at a rate known as the solar constant. This radiant energy is stored for various periods of time and then converted into other forms of energy, but eventually it must all be released from the system to space. There has developed an energy balance within the earth system and between the earth and space, such that over the long run a steady-state condition has evolved; the earth is not getting warmer or colder at a discernable rate. The earth can also be considered a single unified system driven by solar energy, consisting of four main subsystems: the atmosphere—the gaseous layer surrounding the earth; the geosphere—the rocky crust of the earth; the hydrosphere—water in all its forms; and the biosphere—living matter. This separation of the physical environment into four parts is quite arbitrary since there are no clear-cut boundaries between them and there are many interrelationships between the subsystems. This categorization is made for the convenience of selecting areas of the environment for study. This book is concerned primarily with the atmosphere, but the interrelationships that exist between the atmosphere and other areas of the environment should become apparent.

The atmosphere as one of the most powerful influences on human life is a factor in the operation and degree of success of many kinds of economic enterprises; it regulates much of the day to day activity of man; and it is the source of some of the major natural catastrophies experienced by man.

The major occupation of man is agriculture, and the conditions which prevail in the atmosphere largely determine the success or failure of agricultural endeavors. The margin between a good crop and a poor one may depend upon a single rain occurring at a critical period in the growth of a plant species. While an exceedingly wet spring may delay planting, a drought may later destroy it. Violent storms in the atmosphere can completely paralyze modern transportation systems and disrupt communication in general. But far more important is the death and injury to human life and the property damage that these storms sometimes cause. Even if the storms and general capriciousness of the weather are omitted, man still finds himself continually adjusting to the changing atmosphere. Wardrobe, recreational activities, and in many cases work activities change according to weather. Perhaps even man's mental state shifts with the weather.

As time goes on, the ability of man to control his environment is increasing rapidly. The point has been reached in the construction of homes and other structures where the atmosphere inside the structure can be almost completely controlled in terms of light, heat, and moisture. Windows in such buildings serve more as a psychological than as a physical need. Eventually entire communities will be living in artificially controlled environments. Control of the atmosphere extends outside the closed structures as well. Frost and fog dispersal are examples of small scale atmospheric modification, and artificial stimulation of precipitation is an example of modification at a larger scale. Evidence is beginning to accumulate which shows that man may be exerting a far stronger influence on the atmosphere inadvertently than he is able to do deliberately. Local climate is being inadvertently modified by desert irrigation projects, by construction of large reservoirs, by removal of natural vegetation, and perhaps most seriously by using the atmosphere as a sewer.

Much of the ability to forecast weather, to reduce losses due to storms, and to modify the atmospheric environment is a result of the growth in knowledge of the atmosphere that has taken place in the last twenty-five years. World War II saw the advent of air warfare on a large scale and general global war. A great deal of information about the atmosphere was gathered by raw experience. Armed forces were gathering data by the millions at the same time. This data collection was used in making decisions for air strikes, troop movements, and other activities geared to weather conditions. With the launching of Sputnik I, a new era in weather analysis opened. Orbiting satellites now provide pictures of large regions of the atmosphere, sense temperatures, and measure radiation. This information relayed to the earth is invaluable in storm detection and weather forecasting.

This book as an introduction to the nature of the atmosphere is designed to serve the student in order that he may better understand the character of this environmental subsystem which plays such a significant role in man's activities. In a small way it attempts to show how far man as a species has progressed in upsetting the equilibrium of a critical resource.

Contents

Chapter 1

The Nature of
the Atmosphere

The atmosphere in its normal state is colorless, odorless, and taste-less. When some degree of opaqueness, odor, or taste is present, it is due to the introduction of exotic materials either from physical and biological processes of the natural environment or by the action of man. It is with respect to the emission of large volumes of waste materials from machines and industrial processes into the air that a considerable interest in the problems of air pollution exists.

The atmosphere is made up of a mixture of gases. A gas consists of an extremely large number of particles moving at high velocities in random directions. Some of the particles in the atmosphere are single atoms such as argon and helium, and others are molecules consisting of atoms of different elements like common water vapor and carbon dioxide. The temperature of the atmosphere depends on the average velocity of these particles and the pressure of a gas is due to the impact of these particles on a surface. The gases making up the atmosphere can be divided into two categories: the constant gases which are relatively constant by volume, and the variable gases. The constant gases tend to remain in about the same proportion through time and vary little from place to place in the atmosphere. The three most important constant gases are: nitrogen—78% by volume, oxygen—21%, and argon—9.94%. Nitrogen is by far the most abundant of the gases but it is relatively inactive chemically in the atmosphere. Argon is also inactive but it is present in rather small amounts. Oxygen is both present in a large amount and very active in the chemical processes of the physical and biological environment. Many other gases are found in lesser amounts and are constant by volume. Table 1.1 lists the more important constant gases. The constant gases stay fairly constant in proportion up to a height of some 80 kilometers (48 mi).

TABLE 1.1

Chemical Composition of the Atmosphere

Major Constant Constituents	
Substance	**Per Cent by Volume**
Nitrogen (N_2)	78.09
Oxygen (O_2)	20.95
Argon (Ar)	0.93
Carbon dioxide (CO_2)	0.03
Neon (Ne)	1.8×10^{-3}
Helium (He)	5.2×10^{-4}
Krypton (Kr)	1.0×10^{-4}
Hydrogen (H_2)	5.0×10^{-5}
Xenon (X)	8.0×10^{-6}
Ozone (O_3)	1.0×10^{-6}
Major Variable Constituents	
Water, solid, liquid and gaseous	
Solid particles, organic and inorganic	

Data from *Smithsonian Meteorological Tables,* Washington: Smithsonian Institution, 1966.

The variable gases of most importance are water vapor and carbon dioxide. Water vapor content varies from near zero to a maximum of about four per cent by volume. Carbon dioxide is found in amounts that average near 0.03 per cent and is considered by some to be one of the constant gases. In terms of the spatial, or place to place, variation within the atmosphere, some of the minor constant gases and the variable constituents are of more importance than the major elements. Water vapor is the source of all precipitation which falls on the earth. Water vapor, carbon dioxide, and dust all act as absorbers of radiation whereas nitrogen and oxygen, the two most abundant gases, are quite transparent to solar radiation. Ozone is normally found in the atmosphere in small amounts. The average is about one part per million. There is a concentration of ozone from 25 to 30 kilometers (15-18 mi) where it is highly absorbent of ultraviolet radiation. The ozone is formed from the combined impact of molecules of nitrogen, oxygen and atomic oxygen. The atomic oxygen exists as a result of splitting of oxygen molecules by shortwave solar radiation. Occasionally high concentrations of ozone occur in the air over cities as an element of smog. Ozone is a highly poisonous gas as well as being highly corrosive. The ozone is formed in smog by a photochemical reaction on nitrogen dioxide which is a product of the combustion of fossil fuels.

Another variable component of the atmosphere which in many ways acts like a gas is the particulate matter suspended in the air. This includes soil particles, smoke residue, ocean salt, bacteria, seeds, spores, volcanic ash and meteoric particles. The primary source of the solid particles, with the exception of the meteoric dust, is at the earth's surface so that particulate matter decreases rather rapidly away from the surface. The meteoric dust which enters the atmosphere amounts to some five billion kg per year. High altitude particles originate primarily from meteoric dust, severe volcanic eruptions and nuclear explosions. Since the advent of nuclear power in 1945, fine particles such as strontium 90 and cesium 137 have been present in the atmosphere. Particulate content varies from as little as 100 parts/cm^3 to several million parts/cm^3. Water and solid particles are largely responsible for the irregularities that occur in the amount of solar energy received at the surface of the earth, and the solid particles serve an additional function as nuclei for condensation of water particles.

Physical Characteristics of the Atmosphere

Although the atmosphere is a mixture of gases, it acts as a single gas in many ways. The mixture of gases acts according to the general laws that apply to all gases. These gas laws state the relationships that exist between pressure, density, volume, and temperature in gases. Robert Boyle, a British physicist and chemist (1627-1691) developed two of these laws. One of the laws states the relationship between pressure and the volume in a gas. Where P is the pressure of the gas, V is the volume, T is the temperature, and K is a constant:

$$P_0 V_0 = P_0 V_0 = k \quad T = K \tag{1}$$

Boyle's law states that *as pressure increases, the volume will decrease,* due to the compressibility of gases. As the pressure on a given volume of air is decreased the air will expand increasing the volume or space occupied by the given parcel of air. Temperature is assumed constant in this case as changes in temperature affect pressure and volume. Boyle's second law follows essentially from the first, relating pressure and density. Where P is the pressure, D the density, and K a constant:

$$P/D = K, \ T = K \qquad (2)$$

If the volume of a given mass is changed the density changes.

The preceding laws show the relationship between pressure, volume, and density, assuming a constant temperature. Temperatures are continually changing in the atmosphere and these changes in temperature affect the pressure and volume of the gasses as illustrated by Gay-Lussac's law and Charles' law. Gay-Lussac's law states the relationship between volume and temperature. If V_o is the volume at $0°C$,[1] V_t the volume at temperature t, and t is the temperature in $0°C$:

$$V_t = V_0(1 + 1/273), \ P = K \qquad (3)$$

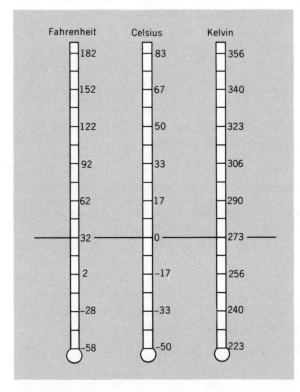

FIGURE 1.1 Temperature scales in most common usage. The Celsius scale is the centigrade scale.

---

1. °C=degrees Celsius, (or centigrade). Since 1948 the International Committenn on Weights and Measures and the National Bureau of Standards has urged that the name Celsius be used. The centigrade thermometer was invented by the Swedish astronomer Anders Celsius (1701-1744) in 1742. In accordance with standard meteorological pracice, °C is used throughout this book. To convert to °F:

$$°F=9/5(°C)+32$$
$$°C=5/9(°F-32)$$

The law shows that there is a direct relationship between volume and temperature. *As temperature increases volume increases,* if the pressure is held constant.

Jacques Charles' law states the interdependence between pressure and temperature. *If the temperature is increased the pressure is increased,* if the volume is held constant. Where P_t is the pressure at temperature t in °C, and P_0 is pressure at 0°K (-237° C):[2]

$$P_t = P_0(1 + P_o/273t), \quad V = K \tag{4}$$

The *combined gas law* places temperature, volume, and pressure into one statement, making use of a constant r, which is constant for a specific gas and varies from gas to gas. T is in °K.

$$PV = rT \tag{5}$$

The atmosphere has not only the characteristics of a gas but also those of a fluid. A fluid is a substance having particles which move easily and change their relative position without separation of the mass, and which easily yield to pressure, and thus are capable of flowing as in plastics, liquids, or gases. Water is also a fluid and in some ways water and the atmosphere act similarly. Motion in fluids can be motion in any direction, depending on the direction of the pressure applied, and the flow can generally be classified as laminar or turbulent. Laminar flow occurs when each individual particle in the fluid travels in a straight line with the path taken by each particle being parallel to the paths

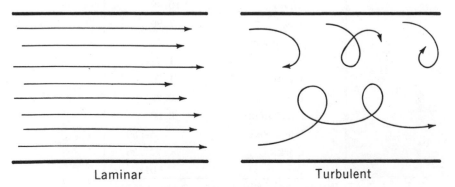

Laminar   Turbulent

FIGURE 1.2 Laminar and turbulent flow. The arrows represent the paths of individual particles in the fluid. Net direction of flow will be the same in both cases.

2. °K degrees Kelvin. Zero °Kelvin is absolute zero, or the temperature at which substances have no heat. °K=°C+273

taken by other particles. Turbulent flow occurs when the individual particles are traveling in most any direction with respect to the mean direction of travel of the fluid. The eddies in streams and the circulation in a dust devil are examples of turbulent flow.

An additional characteristic which is often overlooked is that the atmosphere has mass. As a result it exerts pressure on any boundary surface such as the ground or the human body, and friction is exerted when the atmosphere is in motion or when objects pass through it.

The atmosphere contains energy in a variety of forms. Dry air in the atmosphere contains energy in three forms. Energy exists as geopotential energy which is a function of the gravitational forces of the earth and is proportional to the height of the air above the ground. Energy exists in the air as internal energy due to molecular motion of which temperature is an indicator, and kinetic energy which is a function of wind velocity. The kinetic energy of the atmosphere is eventually dissipated through friction. To the energy of dry air can be added the latent heat of water vapor.

TABLE 1.2

Densities of Selected Substances

Density in grams per cubic centimeter	
Gold	19.30
Mercury	13.59
Silver	10.49
Iron	7.90
Aluminum	2.70
Clay	1.8-2.6
Water	1.00
Ice	0.92
Paraffin oil	0.80
White Pine	.35-.50
Cork	.22-.26
Air	.0013

From *Smithsonian Physical Tables*, 9th revised edition, Washington: Smithsonian Institute, 1964.

Vertical Distribution of Atmospheric Properties

There are marked changes in some aspects of the atmosphere with height. The atmosphere is an unconfined gaseous fluid resting on the surface of the earth. As a result of its having mass and also being compressible, the mass and pressure of the air decrease with height. The portion of the atmosphere closest to the surface is subjected to the mass of gases above, and hence the greatest portion of the mass of the atmosphere lies near the surface. Of the total mass of the atmosphere, approximately 50 per cent lies below 5,500 meters or five and a half kilometers (3.5 mi). Some 84 per cent is below 13 km (8.5 mi). The density of the atmosphere decreases at a geometric rate with height but the gases stay in roughly the same proportion up to heights of at least 80 kilometers (48 mi). The gases continue to decrease in density and continue out for some 1000 km (600 mi).

The atmosphere is divided vertically into a number of different zones, based upon a variety of changes which occur with height. The lower zone is the troposphere. The troposphere is characterized by motion or flow both vertically and horizontally, and by a rather uniform decrease in temperature with height. The zone can be divided into two parts, the lower and the upper troposphere. The lower troposphere which extends upwards of three km above the surface is the zone in which there is maximum friction between the earth and the atmosphere, and it is the area most affected by the diurnal (daily) changes in surface conditions. Another characteristic of this zone is the frequent existence

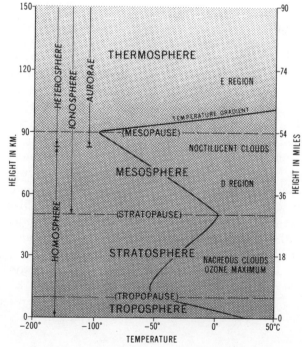

FIGURE 1.3. The vertical structure of the atmosphere. The diagram is generalized as there is some variation in the height of the lower layers with time and space.

of temperature inversions. A temperature inversion is a situation where the temperature increases rather than decreases with height as is the usual case. The upper troposphere which extends to a mean height of 12 km (7 mi) is largely exempt from the diurnal changes which take place near the surface and from the effects of friction at the surface. The primary changes which take place here are associated with the secondary circulation (atmospheric storms) and with the seasons. The upper reaches of the troposphere are frequently marked by a distinct boundary zone known as the tropopause. There are actually a series of overlapping tropopauses existing at different heights. The tropopause is significant in a number of ways. For the most part, it marks the upper limit of surface initiated turbulence; it represents a cold point in the vertical temperature structure of the atmosphere; and it also marks the upper limit of most of the energy contained in the atmosphere in the form of latent heat of evaporation. Most of the mass of the atmosphere lies below the troposphere also so most of the geopotential energy of the atmosphere lies below this height as well. At the equator the tropopause exists at a height of 16-17 km, a temperature of -70 to -85°C (-94 to 120°F), and at an average pressure of only 100 mb (3 in). At the poles, the tropopause is at a height of 9-12 km (5.5-7 mi), a temperature of -50°C to -60°C (-58 to 76°F), and at a pressure of some 250 mb (7.5 in).

Above the tropopauses is the stratosphere, named for the layered nature of the air at these levels. In the stratosphere, the temperature at the top is approximately the same as it is at the bottom although there are a number of marked fluctuations within the zone. The lower point, up to about 20 km (12 mi), is nearly isothermal. The upper region extending to 50 or 60 km (30-36 mi) is characterized by temperature increases with height which range up to as much as 4° C/km (12°/mi). The increasing temperatures in the stratosphere are due to the absorption of ultraviolet radiation. The peak of this photochemical reaction is at 25-30 km (15-18 mi) where the ozone concentration is highest. This ozone layer acts as an upper atmospheric heat source. The energy absorbed here is sent downward in the atmosphere and out into space. The lower stratosphere in fact is a cold zone marking the point between the upward movement of heat by convection from the ground surface and downward radiation of energy from the ionized ozone layer. Above the stratosphere is another boundary layer, the stratopause. As is the case with the tropopause, there may not be a single boundary layer but instead possibly a series of overlapping layers.

The stratosphere merges with the ionosphere around 80 km (48 mi). As the name infers, the ionosphere contains ionized gases resulting from absorption of solar radiation. Above this height the atmospheric tend to be stratified. At heights of some 100 km (60 mi) and 250 km (150 mi) the major ionized layers exist; these layers are of major significance to radio communications. The ionosphere is variously referred to as the thermosphere or chemosphere.

The auroras of the north and south polar regions are produced in the ionosphere. The auroras are formed when excessive numbers of charged particles enter along the outer Van Allen belt. They strike the nitrogen and oxygen giving off colored light and X rays. The color of the aurora depends upon the altitude at which the collisions occur. At about 1000 km (600 mi), the shapeless blue and red glows occur. The blue is produced by ionized nitrogen and the red by ionized oxygen—the latter form being quite rare. At altitudes of 80-300 km (48-180 mi) the yellow occur. This is a mixture of the high level red and the low level greens. Overlapping the yellows in the lower altitudes are the green rainbows and certain forms which develop from 100 to 250 km (60-150 mi). These are formed from charged oxygen. At some 80 km (48 mi), energized nitrogen molecules occasionally produce a scarlet lower rim.

Beyond 500 km (300 mi) is the exosphere. Here initial interaction occurs between the atmosphere and solar radiation. Here also is where the meteorites begin to interact with gaseous matter. The outer edge is a spray region for gaseous particles. Particles are far enough apart so that some which have a high velocity in a direction away from the earth, may escape its gravitational force and escape into space.

The condition of the atmosphere near the surface is affected by the processes which operate as far out into space as the normal atmospheric gases go. The higher in the atmosphere that a change occurs the longer it will be before it is reflected in the atmosphere at the surface. The hourly weather is associated with changes in the lower troposphere, the day to day weather changes are associated with the storms of the secondary circulation, which are at least partially controlled from the upper troposphere. Changes over longer periods of time are associated with changes in the upper atmosphere and are detected in seasonal changes. These changes are brought about by changes in solar activity.

Weather and Climate

Weather is the state of the various atmospheric elements at a particular point in time and space. The elements are continually changing with time so that a major characteristic of weather is a high degree of variability. These changes with time result from the atmosphere being a constantly moving mixture of gases and other substances, with the movement resulting from the external influence of the sun and the earth.

Weather is the primary ingredient in climate. Climate consists of the regime of weather at a place over a relatively long period of time. It includes the most frequently occurring types of weather as well as the infrequent types. The typical weather may be described by average temperatures, precipitation amounts, and the types of storms that are most common. The extremes of weather consist of variations from the normal such as too much or too little rainfall which may produce floods or draughts, high or low temperatures which can cause considerable discomfort as well as property damage, high velocity winds, or some combination of these conditions. No place is free from the effects of some types of severe weather. Periods of severe weather are usually not frequent by definition, but they may result in loss of life and extensive property damage, and as such are an important part of the climate of an area. Often when the climate of a particular place is discussed the mean conditions are overstressed to the point that any concept of variable weather conditions is suppressed. While weather is variable due to the time and place, climate is relatively constant with time but varies with space. Climatology is concerned with study of the similarities and differences in weather conditions as they occur from time to time and from place to place over the surface of the earth.

The analysis of weather and climate is usually centered on several variable characteristics of the atmosphere near the surface: radiation and temperature, humidity and precipitation, and pressure and winds. Each of these elements varies with space and time, and the separation of these elements, like the division of the physical environment, is arbitrary and not without overlap. There are many interrelationships between the variables.

The weather that is experienced at a place is often recognized as being subject to identifiable temporal sequences. Regular changes in the elements often occur daily and annually. These changes are examples of periodic fluctuations in the atmosphere. Cyclic changes in weather are represented by periods of years with unusually large amounts of rainfall followed by irregular periods of drier years. These cyclic events recur but at various intervals. The changes which occur within the atmosphere do not constitute random events which lack systematic order; these weather patterns vary systematically over the surface of the earth in an orderly manner so that there is an identifiable pattern of climatic differences from place to place.

The controls of weather and climate are those factors responsible for, or associated with, the variation of the elements of weather and climate through time and space. The distribution of solar energy through time and space and the rotation of the earth about its axis serve as the major controls of weather and climate. These two factors produce the general circulation patterns of the atmosphere and the primary distribution of temperature over the earth. Other factors that may play a large role locally are the distribution of land and water and topography. The secondary circulation, namely the different types and frequency of storms and shifting local winds, provide the daily and seasonal variety of weather.

Chapter 2

Radiation
and Temperature

There are basically three ways in which heat can be transferred from one place to the other: conduction, convection, and radiation. Although these means of heat transfer are utilized in heating the atmosphere, they are not of equal importance. Conduction is the direct transfer of heat from molecule to molecule, and represents the flow of heat along a temperature gradient. The closer together the molecules are packed and the more contacts between them, the faster the energy will be transferred. Thus, conduction takes place most readily in solids, to a lesser extent in liquids and is least effective as a means of heat transfer in gases. Since the atmosphere is a gas, conduction plays but a small part in the exchange of heat in the atmosphere. Conduction between the ground and the air does occur but this takes place in a very limited zone near the surface.

Convection involves the transfer of heat by means of mass motions of the medium through which the heat is transferred. Thus, instead of having the energy transferred from molecule to molecule, the molecules which are energized are moved bodily from one location to another. This is possible only in fluids (liquids or gases) because they alone have internal mass motion. This does not take place in rigid bodies. In the process of convection the moving masses carry with them heat acquired by conduction or some other means such as radiation or evaporation while in another location. Since the atmosphere is a medium in which mass motions are easily started, convection is the major way in which heat is transferred within the atmosphere. Heat transfer in this fashion can be either vertical, horizontal, or a combination of the two. In reference to the atmosphere the two types of motion are often separated. The vertical motion being referred to as convection and the horizontal motion assuming the title of advection. Complete vertical cir-

TABLE 2.1

Thermal Conductivities of Some Common Substances[1]

Substance	K in Cal/cm. sec. °C	K in Btu/ft. hr. °F
Silver	0.99	242
Copper	0.91	225
Aluminum	0.49	118
Tin	0.15	37
Ice	0.005	1.2
Earth's Crust	0.004	0.96
Dry Soil	0.0037	0.88
Water	0.0015	0.37
Snow	0.0011	0.27
Air	0.000054	0.017

From *Smithsonian Physical Tables*, 9th revised edition, Washington: Smithsonian Institute, 1964.

culation may take place in a single cloud. While not all advection is large scale, certainly the largest circulation cells in the atmosphere are essentially advective and involve heat transfer slowly from the equator to the polar areas. For the earth-atmosphere system advection is the most important means of heat transport. The day-to-day changes in the mid-latitudes are due to advection of different kinds of air into the area. Both natural and mechanical convection and advection occur in the atmosphere. Natural convection develops due to differences in density of the atmosphere and forced convection is the result of some external force.

The Nature of Radiation

Both conduction and convection depend upon the existence of a material medium to operate. This medium may be a solid, liquid, or gas. Radiation is the only means of heat transfer through space without

---

[1]K is the thermal conductivity of the substance. A calorie is the amount of heat needed to raise one gram of water one degree Celsius. There is some variation in the heat necessary to raise the temperature of water a unit amount so specifically it is the amount of heat necessary to raise the temperature from 14.5 to 15.5° C. The thermal conductivity is expressed in calories per second passing through an area of 1 square centimeter when the temperature gradient is 1°C per centimeter. It is also expressed as British Thermal Units (BTU's) per hour transferred through 1 square foot with a temperature gradient of 1°F per foot. A BTU is the amount of heat required to raise 1 pound of water 1°F (from 63° to 64°).

the aid of a material medium. The source for the heat in the atmosphere can be traced indirectly to the sun, and between the sun and the earth, where only a minimum of matter in the form of solar plasma exists; radiation is the only important means of energy transfer.

There are quite a few terms used in connection with radiation that are not frequently encountered elsewhere. A few of these terms and their definitions are included here to facilitate the discussion that follows.

definitions

Radiant Energy - electromagnetic waves which travel at about 310, km per second (186,000 miles a second).

Black Body is a substance which absorbs all radiation which reaches it. While this situation is never actually reached, it is approached by many objects.

Black Body Radiation is the highest theoretical radiation rate that a body can possibly have at a given temperature. Both the earth and the sun approximate black body radiators.

Emissive Power is the energy radiated from a surface per unit of time. The radiation intensity varies as the fourth power of the temperature in degrees Kelvin.

Absorption Rate is the energy absorbed on a surface per unit area per unit time. The absorptivity of a black body is 100 per cent and all other bodies less. All objects absorb and radiate energy. The difference between the amount absorbed and the amount radiated is equal to the heat added or lost by the object.

The sun which is essentially an atomic furnace radiates maximum energy corresponding to a temperature of about 5,700° C (10,000° F) and at a pressure of approximately 100,000 calories per square centimeter per second. Traveling at a rate of speed of 310,000 km per second (186, 000 miles per second), the radiant energy reaches the earth in some six to eight minutes traveling over a mean distance of 155 million km (96 million miles). Of the total energy given off by the sun only about one-half a billionth of it reaches the earth. This energy reaching the earth is some two gram-calories per minute. The amount varies from a high of 2.06 in January when the earth is closest to the sun to a low of 1.94 in July. Since over the long run the amount doesn't vary much, it is referred to as the solar constant. An often used measure of radiation is the Langley which corresponds to one gram-calorie per cm² per minute. Since the solar constant must be spread over the whole earth's surface, the average energy received is one-fourth of the solar constant or one-half Langley. The solar constant thus yields an average daily total of 720 Langleys.

Radiant energy wavelength is measured in either microns or angstroms-one micron equals 1/10,000 cm (1/25,000 in); one angstrom equals 1/100,000,000 cm (1/250,000,000 in) in length.

Radiant energy, radiant heat, and light constitute essentially the same form of energy. The difference between them is man's ability to perceive them. Man's vision responds to a portion of the wavelengths in the electromagnetic spectrum only. If a black body has a temperature above 1,000 degrees Centigrade, (1850° F) some of the radiation emitted by it is in the visible region. About half of the solar radiation is in the visible portion of the electromagnetic spectrum which is from 0.4 to 0.7 microns in length. Wavelengths shorter than 0.4 microns are the ultraviolet radiation and radiation longer than 0.7 is the infrared range.

Effects of the Atmosphere on Incoming Solar Radiation

The solar radiation received at the surface of the earth is less than that emitted by the sun and less than that received at the outer edge of the atmosphere. The solar constant, or energy received at the outer edge of the atmosphere, is 0.50 calories per square centimeter per minute over the whole atmosphere, and at the ground approximately 0.25 calories per square centimeter is received. The solar constant actually changes a few per cent with time due to the change in the distance to the sun and changes in solar activity

Part of the solar energy received at the outer edge of the atmosphere is reflected back to space by clouds, solid particles in the air, and the surface of the earth. About 1/3 of the incoming radiation is reflected back to space, most of this by the atmosphere but a small amount by the surface of the earth. Clouds reflect by far the most radiation for the earth as a unit. The lack of light on a cloudy day is due more to reflection than to absorption, and the thickness of the cloud layer is the key to the amount reflected. Reflection does not affect color, only light intensity. All wavelengths are reflected equally so there is no separation of color.

A second way which incoming solar radiation is affected by the atmosphere is by scattering. When radiant energy passes into a transparent medium containing small particles or molecules, some of the radiation is deflected in various directions. This scattering, known as Rayleigh scattering, occurs only when the diameters of the particles are

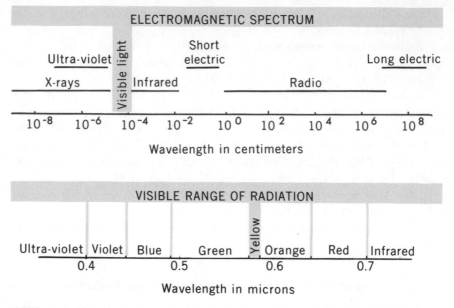

Wavelength in microns

FIGURE 2.1 The electromagnetic spectrum. The visible range of radiation is an expansion of that portion of the electromagnetic spectrum lying between 0.4 and 0.7 microns in length.

smaller than the wavelength of the radiation. Particles of a size larger than the wavelength do not produce scattering of this kind. The scattering process in the atmosphere is selective, the shorter wavelengths such as blue are the most effected of those in the visible range. The blue color of the sky is the consequence of a more complete scattering, primarily by the stable constituents of the atmosphere. After a rain or during a period of dry calm weather the sky appears as a much darker blue than when there is dust or abundant moisture in the air. The less scattering, the darker blue the sky. In ascending a mountain or in an airplane the difference in sky color is very noticeable. In the evening when the sun's rays pass through the greater length of the atmosphere, the red portion of the spectrum becomes visible as the blue sector is completely diffused or scattered out. The scattering of light waves is the main cause of reduced visibility over great distance. Under hazy or dusty conditions the light from a distant object can be completely lost through diffusion. About a fourth of the incoming radiation is scattered, but much eventually reaches the ground.

A third form of depletion of solar energy is by the direct absorption by the atmosphere. The atmosphere is absorbant in only some parts of the electromagnetic spectrum. Much of the energy absorbed by the atmosphere is absorbed in short wavelengths high in the atmosphere. Ultraviolet radiation with wavelengths of less than 0.3 microns are absorbed by the ionosphere in the layer where ozone is at its maximum level near the stratopause. On the other end of the scale water vapor and carbon dioxide absorb in the range from 0.7 to 16 microns or within the infrared range. Thus, the atmosphere is fairly transparent to radiation in the band of visible light where solar radiation is the strongest. Only when the variable components of the atmosphere are present in large amounts near the earth's surface is atmospheric absorption of solar radiation significant.

Some materials will allow radiation to pass through unaltered and such a material is referred to as being transparent to the radiation. Not only does radiant energy pass through space without the presence of a large amount of matter but it will also pass through some kinds of matter such as water, glass, and certain gases. In this process of transmission, the radiation itself is not affected and the material through which it passes is not altered. The solar radiation that is not reflected, scattered, or absorbed is transmitted to the surface of the earth. The depletion of solar radiation entering the atmosphere depends on the length of the path through the atmosphere and the amount of the variable constituents which are present. Some 33 per cent of incoming radiation is transmitted through the atmosphere to the earth essentially unaltered.

The amount of radiation reaching the atmosphere is fairly constant. If there is a change in the amount of radiation reflected there must be a change either in the amount absorbed or transmitted or both.

The Earth's Surface and Radiation

The amount of solar radiation which reaches the ground surface via transmission and scattering is referred to as insolation. The earth surface as a whole is not very reflective and the average amount reflected (the albedo) is between five and ten per cent of the energy which reaches the surface. The reflectivity is determined mainly by the angle of incidence, the color, and texture of the surface. Dark colors and rough surfaces absorb more radiation than light colors and smooth surfaces. Radiation received by soil is changed to sensible heat and retained near the surface. Soil does not have a high specific heat and does not serve as a good conductor of heat. In a dry sandy soil, heat received in the daytime will be radiated away at night, which results in large diurnal contrasts of soil temperatures. The more moist the soil is the higher the specific heat and lower the diurnal range of temperature. The lower few meters of air reflect the temperature regime of the surface. Annual variability in soil temperatures is not discernible in mid-latitudes to depths greater than 80 meters (260 ft). Water bodies respond to solar radiation very differently than land masses even though the amount received at the surface is about the same. Water reacts differently for several reasons: (1) water has a very high specific heat, and is in fact used as the standard for measuring specific heat—the specific heat of water is some 3 times greater than earth materials; (2) some 25-30 per cent of the energy received at the surface is used to evaporate water; (3) since water is a fluid the energy received at the surface is distributed both horizontally and vertically by internal motion. The energy is thus not concentrated in the surface layer to the extent that it is on a land surface; (4) the energy is transmitted to greater depths in water. Clear water is relatively transparent to solar

TABLE 2.2

Specific Heat of Some Common Substances[2]

Substance	Specific Heat
Water	1.00
Ice	0.50 (at freezing)
Air	0.24
Sandstone	0.22
Limestone	0.21
Aluminum	0.21
Granite	0.19
Sand	0.19
Iron	0.11
Silver	0.06
Mercury	0.03

From *Smithsonian Physical Tables*, 9th revised edition, Washington: Smithsonian Institute, 1964.

---

[2]Specific heat is the heat capacity per unit mass of a substance. It is the amount of heat required to raise the temperature of the substance by 1°. It is expressed in calories per gram per degree Celsius, or BTU's per pound per degree Fahrenheit. The value of the specific heat is the same in both units.

radiation. There is a practical limit to which the radiation will go due to dissolved salts in the ocean and sediment in lakes and streams. Whereas annual changes in temperature on land go only a few tens of feet into the soil in the oceans annual temperature changes are detected to depths of 3000 meters (10,000 ft) on occasion. Molecular conductivity of water is actually less than that for dry soil so conductivity does not play a major role in the transfer of heat in the ocean. The transfer of heat in the ocean by turbulence probably is greater than that by conduction by a power of four or five times.

The ocean is a tremendous reservoir of heat, absorbing energy in summer and releasing it in the winter. The ocean in tropical areas absorbs radiation at 100 to 150 Langleys per day. The ocean currents driven by the general circulation carry the heat poleward where large amounts are released into the air by currents such as the Gulf Stream and Kyushu Current. In latitudes as high as 45°, the heat release may exceed 200 Langleys per day. When cold air flows over the warm water energy transmission goes over 1000 Langleys per day. The same kind of phenomenon occurs over large lakes.

The earth disposes of the solar energy it has received by evaporation at the surface, conduction at the interface, and by radiation. The energy that is placed into the lower atmosphere either as latent heat or sensible heat is distributed upward by convection. Where the lower atmosphere is heated intensively, thermal currents develop that carry the heat aloft fairly rapidly. This form of heat distribution will take place upwards for several thousand meters but will usually terminate at the base level of clouds if they are present. Over deserts these thermal currents have been detected as high as 15,000 meters (50,000 ft).

The earth receives solar radiant energy at the rate of 0.25 calories per unit area. If it radiated an equivalent amount it would indicate a mean temperature for the earth of -40 degrees C (-40° F), which is obviously much colder than the existing temperature. As the actual mean temperature for the earth is near 15° Centigrade (59° F), there is an additional supply of heat, or an accumulation of heat.

Earth radiation ranges from 5 microns to 50 microns in length. The maximum is at about 10-14 microns in length and the energy rate is from 0.1 to 1.0 calories per square centimeter per minute. The amount of earth radiation at any time is dependent upon the surface temperature, and this varies widely over the earth's surface. A wavelength of ten microns corresponds to a mean temperature of 15° Centigrade (59° F) and a perfect black body radiator would radiate at 10-15 microns at this temperature. As the sun has its maximum output at about

0.5 microns in length the earth radiation has a mean wavelength of around 20 times that of solar radiation, hence the distinction between the longwave earth radiation and shortwave solar radiation.

The atmosphere is fairly transparent to solar radiation but much less transparent to earth radiation. The stable gases are quite transparent to earth radiation but the variable gases, particularly water vapor and carbon dioxide are not. The amount of water vapor or condensed water droplets in the atmosphere is the major factor in determining the amount of absorption of earth radiation by the atmosphere, just as they are in determining the amount of solar radiation absorbed by the atmosphere. Both total amount of water vapor and its height above the ground affect the absorption. Total vapor content or precipitable moisture varies from near zero to two or three inches (7 cm) in warm moist air. Water vapor absorbs best at wavelengths between 5.7 and 7 micons in length and again at wavelengths in excess of 27 microns in length. Thus the main wavelengths for radiation to escape from the earth are between 8 and 13 microns in length. This is in the infrared band. The energy absorbed by the atmosphere is reradiated in all directions, part of it back to the earth. Reradiation or the return of radiation to the earth is greatest when the moisture is concentrated in the lower few thousand feet of the atmosphere.

Clouds are major absorbers of earth radiation, and they also act as black body radiators. They have a net gain of heat at the base and loss at the top. When clouds exist over an area the radiation surface is moved upward from the ground to the top of the cloud surface. Clouds function two ways to modify earth temperature. They reflect solar radiation away from the surface, thus reducing incoming radiation, and they absorb earth radiation and reradiate it back, thus keeping heat in the atmosphere near the ground. This trapping of solar radiation is known as the greenhouse effect. The clouds tend to produce more modrate temperatures at the surface by reducing heat influx during the day and retarding loss at night. Clouds tend to reduce extremes of temperature but probably have little effect on mean temperatures. Diurnal temperature ranges are essentially a function of cloud cover and water content of the atmosphere. It should be noted too that in winter in mid-latitudes and in the polar areas clouds serve as heat sources for surface areas just as they operate at night as energy sources.

The Heat Balance of the Atmosphere

For periods of time numbering in the tens or perhaps hundreds of years it can be assumed that the earth is becoming neither warmer nor cooler. Since the major source for heating the earth is the radiation from the sun, as radiation is the major means of transferring heat though space, the earth must be losing energy via radiation at the same rate it is receiving it.

Since the solar constant is 720 Langleys per day the outgoing planetary radiation must be equal. There exists a complex mechanism of energy cycling on the earth that eventually gives rise to the balance. Some 35 per cent of the solar radiation is directly reflected back to space. It has already been stated that the average albedo of the earth's surface is around 10 per cent. If the planetary albedo is 35 per cent then a substantial portion of the total reflected radiation must be reflected by the atmosphere, namely from clouds. Cloud formations normally intercept about 40 per cent of the total solar beam. Cloud reflectivity ranges upwards to 75 or 80 per cent. The average reflectivity is 60 per cent of the intercepted radiation. Of the rest of the intercepted radiation five per cent is absorbed and the rest transmitted through the cloud. Thus, the cloud formations with an albedo of 60 per cent are the major planetary reflectors of solar radiation. Some of the measured planetary radiation is scattered radiation that goes back to space.

Stefan-Boltzman's law states that the amount of radiation from a black body is proportional to the fourth power of the absolute temperature. The maximum concentration of radiation from the surface of the earth is at about 10 microns corresponding to a mean temperature of 15° C (59° F). The rate of radiation is greater than the solar constant by some 15 per cent. Since a considerable amount of the solar beam is reflected back to space there must be a secondary source of energy at the earth's surface. This secondary source of energy is the counterradiation of longwave sky radiation. The 65 per cent of the solar constant that is radiated back to space represents a radiating surface with a mean temperature of -23° C (-9° F). Since the earth surface has a mean temperature of 15° C (59° F) then much of the actual loss to space must take place from the cloud surfaces and water vapor high in the atmosphere. Without the atmosphere and its built-in energy storing mechanism the earth would have a mean temperature of slightly below freezing. Diurnal temperatures would fluctuate between +100° C and −100° C (+212° F and −150° F) without the atmosphere. The radiation to space can take place from either the atmosphere or the ground surface. When skies are clear the atmosphere is quite transparent to the solar radiation and relatively so for the longwave radiation. Thus when the atmosphere is low in water vapor, CO_2, and solid particles, there is a high rate of solar radiation received at the surface and also of earth radiation back to space. Desert areas in general, and polar deserts in particular represent major radiation windows of the earth.

When water vapor and other variable constituents are present the zone of net radiation loss is shifted from the ground upward to the cloud top or to the top of the moisture layer. Thus the presence of water vapor adds a loop into the heat balance by absorbing and reradiating heat to the surface as well as to space. This loop or exchange of radiation between clouds and the ground moderates the rate of en-

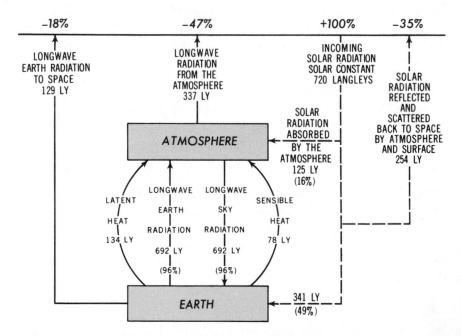

FIGURE 2.2. Radiation balance of the Earth. Dashed lines indicate shortwave radiation, solid lines longwave radiation or other forms of heat transfer. Thirty-five per cent of planetary radiation is in the shortwave bands and sixty-five per cent in longwave bands. (Data from London, 1957.)

ergy receipt and loss at the earth surface and in the lower atmosphere. The more water present in the lower atmosphere the greater the modification of the space to earth to space radiation balance. As moisture content increases reflection from the atmosphere increases, absorption and reradiation by the atmosphere increases, and direct absorption and reflection of solar radiation by the earth decreases.

Spatial Distribution of Temperatures

The amount of heat present is one of the most important variables associated with the atmosphere. Temperature is an indication of the amount of kinetic energy present in the air. The range of temperatures that have been recorded in the lower troposphere ranges from a high of 58° C (136°F) degrees at San Luis, Mexico to a low of −87° C (−125° F) degrees at Vostov, Antarctica. Annual mean temperatures range from a high of 31° C (88° F) at Massawa, Eritrea and Lugh Ferrandi, Somali Republic to a low of −57.8° C (−72° F) at the cold Pole, Antarctica. These data represent the extremes of the conditions that exist, and these places as well as all places are subject to a wide range in actual temperatures. Temperatures vary widely from place to place over the earth's surface as well as through time at an given place.

The most important factor in determining the temperature of the air at a given point over the earth's surface is the intensity and duration of the radiation at the ground surface. The intensity of the solar radiation is largely a function of the angle of the solar radiation with the surface and the duration of radiation.

Latitude is the primary independent variable in determining the spatial variation in insolation on the earth's surface. The differences which occur longitudinally are minor compared to those existing between the equator and the poles. There are two main reasons for the latitudinal variation which occurs and these are the angle of incidence (angle of the sun's rays from the vertical) and the length of the daylight period.

The variation in these values is due to a combination of four factors. They are:

1. The sphericity of the earth.
2. The inclination of the earth on its axis.
3. The rotation of the earth on its axis.
4. The revolution of the earth about the sun.

The more directly the rays of the sun strike the surface the greater will be the energy received per unit area of the surface and the greater will be the amount of energy absorbed. Energy input at the surface will be at a maximum at the latitude where the rays of the sun are perpendicular to the surface of the earth. The portion of the earth lying between latitudes 23 1/2 degrees north and 23 1/2 degrees south experiences perpendicular rays of the sun sometime during the year. The equator experiences the least variation of the angle of incidence of any location on the earth's surface. While the sun migrates from tropic to tropic through an angular distance of 47 degrees the angle of incidence is never more than 23 1/2 degrees. For all locations between the equator and the tropics the angle of incidence varies from 23 1/2 degrees to 47 degrees. The maximum angle is reached at the tropic. From the tropics to the arctic circle the angle of incidence changes 47 degrees through the year.

From the arctic circle to the pole the minimum angle of incidence increases from 47 degrees to 66 1/2 degrees and the seasonal variation decreases from 47 degrees to 23 1/2 degrees at the pole. The two poles are influenced by a seasonal change in angle of incidence comparable to that of the equator, but the angles themselves are much different.

The net result of the variation in the length of day and angle of incidence is that solar radiation is greatest in the low latitudes over the annual period, and at all latitudes it is greatest in the summer season or high sun season, and greatest at noon solar time.

The second factor in determining the amount of radiation reaching the surface is the amount of depletion of the incoming radiation by reflection, scattering, and absorption, which is largely a result of the concentration of the variable gases and solid particles in the atmosphere and the length of the radiation path through the atmosphere. When the sun's rays are perpendicular to the surface the distance which the radiation must travel through the atmosphere is shortest or at a minimum. The shorter the distance the radiation must travel through the atmosphere, the less reflection, scattering, and absortion there will be. The greater the distance the rays travel through the atmosphere, the more radiation will be depleted by the atmosphere.

The second factor affecting the temperature is the spatial arrangement of land and water. Due to the high specific heat of water, and the distribution of heat in a water body to considerable depths, a site which is located near water will experience a more reduced range of temperatures than one at a relative distance. Since the atmosphere is heated primarily from the earth's surface the air overlying a water surface is modified rapidly to develop temperatures characteristic of the water beneath. It is largely this latter aspect which produces longitudinal variations in temperature which are of a seasonal or annual duration. Ocean currents which often carry water of different temperature characteristics into an area have pronounced effects on local air temperatures. Currents of water warmer than is normal for the latitude will warm the overlying air producing air temperatures warmer than is normal for the latitude. By the same token cold currents will lower air temperatures in the vicinity of the current.

The advection of cold and warm air affects local temperatures. Rapid changes in the local energy balance, and other short term changes in temperature, are normally the result of the inflow of air from a region with marked differences in temperature or moisture, or both. Cold waves, heat waves, and the January thaws are examples.

Altitude has a marked effect on temperatures. The change in temperatures with height (the lapse rate) averages about 1° C for each 100 meters (3.5° F/1000 ft) in the lower troposphere. This change in temperature with altitude takes place much more rapidly than the change in temperature with latitude. The mean temperature changes about as much with 1 unit of altitude as with 1000 units of latitude.

In some areas topographic barriers have a significant effect on the temperature of adjacent lowlands. Where such barriers lie in an east-west direction they often serve to reduce the north-south movement of winds. The region around the Mediterranean Sea is protected to a large extent from polar air by the Alps Mountains. While topographic barriers will affect airflow at all latitudes the effects are more pronounced at mid-latitudes where the temperature gradient is the steepest.

Periodic Temperature Changes

An individual residing in one location on the surface of the earth will be able to observe the temporal rather than the spatial pattern of temperatures. There are two major periodic changes in temperature which occur at most places. The first is the annual pattern of temperature change which is largely determined by the factors discussed in the previous section. The second periodic temperature change is the diurnal change. Most sites on the earth's surface experience some diurnal fluctuations in temperature. The degree of the diurnal change is largely dependent upon the variation in the length of day and solar intensity, the concentration of variable gases and dust in the air, and proximity to water.

Both the annual and diurnal temperature regimes are characterized by marked lags. The peak period of insolation in the northern hemisphere occurs at the time of the summer solstice on June 21. This is the time of year when the angle of inclination of the sun is at a maximum and the days are longest. However, the highest temperatures for most areas occur from one to three months after this time. The reason for this lies in the characteristics of black-body radiation. A body radiates heat in proportion to its temperature. As the solar radiation increases during the spring the earth warms up as a result of absorbing this energy. As the earth then increases in temperature the rate of earth radiation increases. During the spring months the rate of solar radiation is greater than the earth radiation and the earth continues to warm. Although the peak of solar radiation occurs on June 21, the rate of solar radiation continues to exceed that of the earth radiation for some time. The maximum temperature comes at that point in time when the rate of earth radiation just catches up to the decreasing rate of solar radiation. From that point on the earth is radiating away energy faster than it is received and so it begins to cool. The diurnal temperature lag is a result of essentially the same process. The peak afternoon temperatures occur up to several hours after the solar noon.

Temperature Inversions

In the troposphere the temperature normally decreases with height as does atmospheric density. This normal decrease of temperature with height averages three to five degrees C/km (two to three degrees/1000 ft) in the lower troposphere and increases to six or seven degrees C/km (three to four degrees/1000 ft) in the middle troposphere. Under certain conditions the vertical temperature regime will be reversed and temperature will increase with height for some distance. With an increase in temperature with height is associated increased density differences and hence a tendency for the atmosphere to be quite stable as the heavier fluid is at the bottom. A temperature inversion represents a reversal of heat flow in the atmosphere. Instead of heat flowing away from the earth it flows towards the earth.

Several different types of inversions occur. Ground inversions occur by cooling from the ground surface. These form when the ground surface cools quickly or over snow and ice and will reach to heights of as much as 1000 meters. They are common in polar regions throughout the year and occur at almost all latitudes periodically. Under clear skies and calm conditions the ground surface will cool more rapidly at night than will the air above it. The layer of air close to the ground will then be cooled and its density increased. Being a fluid, the cooler, more dense air will begin to flow downhill. In the evening cool air is often noticed when descending into a draw or valley. This phenomenon is even noticeable in city streets on occasion where they act as stream beds for the cooler, heavier air. Prerequisite to the formation of these inversions is little turbulence and a clear atmosphere. Whenever there is pronounced vertical motion or convection the inversions will not develop due to the mixing of the warm and cool air. This situation also demands a fairly clear atmosphere. The first indication that such an inversion is developing is indicated by a heavy dew early in the evening. Ground fogs may result, developing in the low areas first and then spreading outward as cooling continues. Inversions of this type are usually local in character.

The second manner in which inversions develop is by slow advection of a more dense current of air into a region over the surface air, or the advection of a warm air mass over a cold surface. These two situations are somewhat different from each other. One produces a very stable atmospheric condition and the other a very unstable situation. Where an air mass moves in over a cool land area the chilling which occurs increases the density of the air at the surface and enhances stability. Where a cool dense air mass moves over the top of a warm dense air mass the lower air may be trapped until some turbulence develops. This kind of movement is most often associated with lee areas near mountain ranges and this type of situation is largely responsible for the smog problems of the Los Angeles area and similar regions.

In the tropics, inversions develop associated with the trade winds that often are lifted to 3000 meters (10,000 ft) with the base anywhere from sea level to 2000 meters (6600 ft). Two major temperature inversions occur aloft, one at the tropopause and the other at the mesopause. These inversions are largely the product of radiation absorption by the upper atmosphere.

Summary

Many factors affect the temperature of the air at a given time. The basic factors associated with temperature conditions are the following:

1. The intensity and daily duration of solar radiation received at the outer edge of the atmosphere. These are largely responsible for the latitudinal differences in temperature.
2. The depletion of this energy by reflection, scattering, and absorption by the atmosphere. The extent of this depletion is largely responsible for arid areas developing higher temperatures than humid areas under similar radiation conditions.
3. The reflectivity of the surface. Areas with light colored surface materials exposed will have slightly lower temperatures associated with them than dark colored materials.
4. Land and water association. Air temperatures over and near water bodies are less subject to extremes than those over land regions.
5. Local heat budget. The moisture conditions of the atmosphere control the diurnal fluctuations of temperature.

6. Advection of air and water. This process largely accounts for the difference in temperatures experienced along windward and leeward coasts in mid-latitudes.
7. Convection of heat by turbulence. This process prevents overheating of the surface. Dust devils and other forms of warm weather turbulence are manifestations of the process.

Chapter 3

Moisture in
the Atmosphere

Water exists on the earth in very large amounts and in all three basic forms of matter. It exists in the solid state as snow and ice, in the liquid state that is most commonly associated with water, and in the gaseous state as water vapor. The physical environment of the earth is such that water is continually changing from one form to another and at the same time moving from one place to another. There are several complex processes involved in changing the state of water, the more important ones being evaporation, condensation, and precipitation. This continual circulation of water constitutes the hydrologic cycle.

Water also occurs in the atmosphere in all three physical states and, of the many gases that make up the atmosphere, water vapor is one of the most important. It is present in relatively small amounts, not generally exceeding 4 per cent by volume, but the significance of the vapor is far greater than the relative amount would indicate. Water vapor is of prime importance as an absorber of radiant energy, thus being a significant factor in the heat balance of the earth. In addition, it is a prime mover of heat through the atmosphere, an indicator of potential precipitation amounts, and a key factor in the physical comfort of human beings and many species of fauna.

The ocean is the direct source for virtually all water found in the atmosphere and on the land masses, regardless of its physical state. About two-thirds of the earth's surface is covered by water, which in itself would be sufficient evidence that the oceans are the major source of water in the atmosphere. In addition, the rate of evaporation from a water surface is much greater than from a land surface. For the United States probably 80 or 90 per cent of the precipitation that occurs had its origin in the ocean no more than two or three days prior to precipitation.

There are two characteristics which indicate the relative importance of evaporation and condensation with respect to the oceans and land masses. Evaporation must exceed precipitation over the oceans to allow a surplus of moisture to be advected over the land masses. If precipitation and evaporation were equal in net amounts over the oceans then there would not be any moisture left in the air to fall as precipitation over the continents. It logically follows then that over the land masses of the earth the net amount of precipitation must exceed evaporation. If this were not the case there would be no rivers to carry the excess back to the ocean. The foregoing statement applies only when the land masses are considered in total. There are obviously many areas where evaporation exceeds precipitation, but these are for smaller regions, just as there may well be parts of the oceans where precipitation exceeds evaporation.

Evaporation

Evaporation is the process by which water changes from a liquid state into a gaseous one. Water is sufficiently volatile in solid and liquid states to pass directly into the gaseous state at most temperatures. The process of change from water to water vapor is called evaporation, and from ice to water vapor is called sublimation. Since the source of water vapor is at the earth's surface, the amount of water vapor present in the atmosphere decreases with height. Most of the moisture is found below 10,000 meters.

definitions

Vapor Tension (or vapor pressure) is the pressure of molecules escaping from the water surface. It is more accurately the pressure of a greater number of molecules leaving the water surface than those returning. This pressure is dependent on temperature.

Partial Pressure. Each gas in a mixture exerts a pressure which is proportional to the product of the percentage of the given gas by volume and the total pressure of the mixture.

Vapor Pressure is the partial pressure due to the water vapor in the air. For the air temperatures that exist on the earth the vapor pressure varies from less than 0.10 mb (millibar) to over 100 mb.[1]

Saturation Vapor Pressure is the pressure of the vapor when it is in equilibrium with a water surface at the same temperature. The saturation vapor pressure is dependent upon temperature.

---

[1] A standard atmosphere exerts a sea-level pressure of 14.7 pounds per square inch. This is equivalent to the pressure exerted by a column of mercury 29.92 inches, or 760 millimeters in height. This pressure is also expressed in millibars, which is another measure of force. A standard atmosphere equals 1013 millibars. A millibar is equal to 1000 dynes and a dyne is the amount of force needed to accelerate 1 gram mass 1 centimeter per second.

Absolute humidity is the ratio of the mass or weight of water vapor per unit volume of air. This is usually expressed in grams per cubic meter but is also expressed as grains per cubic foot.

Specific humidity is the ratio of the mass or weight of water vapor to the total mass of a unit of air including the water vapor; usually expressed as grams of water per kilogram of wet air.

Mixing ratio is the ratio of the mass of water vapor to the mass of a unit of dry air. It is expressed as mass per kilogram of dry air.

Relative humidity is perhaps the most widely known measure of atmospheric humidity. This is the indicator of atmospheric humidity quoted by most forms of mass communication. It is the ratio of the amount of water vapor present in the air, or vapor pressure, to the amount of water vapor the air can hold, or the saturated vapor pressure.

Dew point. The dew point is the temperature at which saturation would be reached if the air mass were cooled at constant pressure without altering the amount of water vapor present.

The process of change of state of water from solid to liquid to gas, or the reverse, involves a substantial energy transfer to or from the surrounding media. The latent heat of fusion is 80 calories per gram. This is the amount of heat absorbed or released by the freezing and thawing of water. The latent heat of evaporation is 600 calories per gram at $0°$ C, which is much greater than the latent heat of fusion. Evaporation accounts for about two-thirds of the heat exchange between the earth surface and the atmosphere. To maintain the present balance of temperature, about 90 cm (35 inches) of evaporation and condensation must occur each year over the entire surface.

The amount of water which evaporates from a given water surface in a given time depends upon a number of factors. They are:

1. The vapor pressure of the water surface. This is directly related to water temperature. The higher the water temperature the greater the surface vapor pressure. When water temperatures are greater than the air temperature, evaporation will always take place.
2. The vapor pressure of the air. The greater the vapor pressure of the air the less evaporation there will be. The rate of evaporation varies directly with the difference between the vapor pressure of the water surface and the vapor pressure of the air.
3. Air movement. Air movement is usually turbulent, which results in the removal of moist air from near the water surface and replaces it with dry air from above. Evaporation thus varies directly with the velocity of the wind, the higher the wind velocity the greater the evaporation.

4. Salinity of the water. Dissolved material reduces the rate of evaporation. The greater the salinity the lower the rate of evaporation. The rate of evaporation from sea water is about 5 per cent lower than fresh water.
5. Depth of the water. In deep water bodies the evaporation regime is substantially different from that of a shallow stream or pond.

Condensation on Solid Surfaces

The reverse process of evaporation is condensation. Condensation occurs when water vapor changes to a liquid with the accompanying energy loss. Condensation on solid surfaces is the result of radiation cooling. Solids cool more rapidly than gases, thus the temperature equivalent to the dew point of the air is reached by solid objects more quickly than by the atmosphere. The water vapor and air coming in contact with the surface is chilled to the dew point and the moisture condenses on the surface releasing heat to it. Condensation on solid surfaces, particularly on the ground surface, is best developed under dry still air, when surface turbulence is at a minimum.

Dew is the most common example of condensation on solid surfaces. Dew forms when the dew point of the air is above freezing. There is cooling at the surface by conduction in sufficient amount to bring the air in contact to the dew point, but the temperature of the air not in contact with the ground surface is still above the dew point. Dew can form at any temperature above the freezing mark.

Frost is a form of condensation on solid surfaces when the dew point is below freezing. In the formation of frost, moisture sublimates as a solid on a solid, passing directly from a gas to a solid and frost can form at any temperature below freezing. Black frosts or dry frosts are not true frosts. Black frosts are referred to when the temperature drops to the vicinity of freezing and there is damage to the plant tissues but the dew point is not reached to produce condensation. These occur when the dew point is too low for condensation to begin to form and thus there is no true condensation of moisture.

Condensation in the Atmosphere

Water vapor does not condense readily without some solid surface to act as a nucleus. While condensation will occur without solid material it demands sub-dew-point temperatures in order to do so. The atmosphere normally contains hygroscopic particles in sufficient amounts to provide a base for condensation. When these hygroscopic particles are not present the relative humidity can be increased to several hundred per cent before condensation will begin. The hygroscopic nuclei vary a great deal in size. The smallest are represented by ions and the largest are other raindrops.

The degree of saturation (relative humidity) at which condensation will occur varies with the size, number, and kind of nuclei present. The larger the particles, the more abundant the particles, and the more absorbent the particles, the more readily condensation will occur, and the lower the relative humidity necessary to produce condensation. Most condensation in the atmosphere takes place on the larger nuclei, which are often the more absorbent also, such as smoke, dust, and salt. The larger nuclei, though fewer in number than the smaller, are extremely important, because by the time relative humidity reaches 100 per cent the particles are already the size of fine drizzle and hence of a size sufficient to precipitate in fairly calm conditions. Condensation itself may begin at levels of 80 or 85 per cent relative humidity. Condensation is a process of continuous growth or accretion, as initial condensation may be in terms of particles microscopic in size. Condensation above the ground can be produced by advective cooling, radiation cooling, and adiabatic cooling (cooling by expansion).

Fog is somewhat difficult to define, but it is essentially a stratus cloud that forms at the ground or so close to the ground that it affects the surface visibility. Stratus clouds are characteristic of the lower part of the atmosphere when and where temperature inversions or nearly isothermal layers exist. These generally occur when the air is warmer than the ground surface. The study of fog then involves determining the circumstance under which cooling of air masses with a high moisture content can take place at the surface. (An exception is steam fogs associated with cold weather.) Fog will form in either one of two ways: by cooling the air to the dew point or by the addition of water vapor until saturation is reached.

There are basically two kinds of fog: radiation fog and advective fog. Radiation fogs are tropical or summer phenomena in typical mid-latitude situations and result from cooling of moist stagnant air by radiation.[2] The advective fogs are winter or polar phenomena primarily, and are a product of cooling of moist air passing over a cold surface. When the air is completely still the fog will be patchy and low. A slight air movement will mix the air enough to distribute the fog more evenly over the surface.

---

[2]In the subtropics (such as in Florida) the advent of the dry season permits excessive radiation during clear, calm autumn nights and radiation fogs are quite prevalent in November.

Condensation in the atmosphere near the ground surface is usually a product of cooling by radiation and conduction, but the primary cause of condensation aloft is adiabatic cooling. When a parcel or column of air is lifted in the atmosphere it expands due to the reduction of pressure on it from the surrounding air. The expansion is accompanied by cooling of the air and the cooling takes place at a fairly fixed rate. The temperature in a mass of unsaturated air when lifted through the lower atmosphere cools adiabatically at a rate of 1° C for each 100 meters (5.5° F per 1000 feet) it rises. If unsaturated air is lifted far enough the air will be cooled to the dew point and condensation in the form of clouds will take place. Saturated air, if lifted, will cool adiabatically just as unsaturated air will, but condensation will release the latent heat of evaporation and the net rate of cooling of the rising air will be reduced below that of unsaturated air. The average wet adiabatic rate is 0.6° C for each 100 meters (3.3° F per 1000 feet) of rise, though the actual rate depends upon the amount of condensation which takes place.

Air rising through the atmosphere can become either unstable or stable depending on the temperature and density characteristics of the rising air and the surrounding air. The unstable or buoyant condition occurs when the density of the rising air is less than the surrounding air. If the rising column of air is less dense than the surrounding air it will rise further, just as an ordinary cork will rise to the top of a container of water if it is placed at the bottom of the container and released. If the density of the rising air is the same as that of the surrounding air then the air will cease to rise or become stable. The notion of stability and instability is an indication of the tendency for vertical motion to exist within the atmosphere. Expressed another way: whenever the lapse rate exceeds the dry adiabatic rate (1° C per 100 m) the air is unstable. The lapse rate cannot exceed 3.5° per 100 m. (19° F per 1000 ft). When the lapse rate reaches this value the air will be autoconvective. It will be less dense than that above it and vertical motion will start in the air without any other impetus.

clouds

Clouds are suspended water droplets. In a cloud where turbulence is usually present droplets must be continually forming in one part and evaporating in another. In a small cumulus cloud it is likely that a droplet will remain in existence for only a matter of minutes. In a column of slowly rising air such as in stratiform clouds, individual cloud particles may exist as such for an hour or longer. The development of rain in clouds appears to result from accretion in most cases. The larger the nuclei and subsequent condensation droplets grow the more droplets they intercept until they reach raindrop size. This process is favored by thick clouds and sustained updrafts. The drop must grow quite large before the mass of the particle is great enough to allow it to leave the cloud.

The moisture in the air is quite often supercooled (remains liquid at temperatures well below freezing). Water in liquid form has been detected at $-40°$ C ($-40°$ F). The smaller the particle the lower the temperature may be before freezing. Supercooling takes place most readily in an atmosphere in which there is little turbulence. When supercooling has occurred in the atmosphere some disturbance will cause almost instantaneous freezing. An example of a disturbance would be an airplane in flight. The crystallization of ice does not take place very readily in clouds. It demands supersaturation of approximately 10 per cent. Once begun the process of ice formation is self-perpetuating.

The classification of clouds now in use is essentially an international one. It consists of four families with ten main types. The system can be carried much further to include as many types of forms as desired. The classification recognizes over 100 different forms.

Stratus and cumulus clouds are distinguished on the basis of the mechanism of lifting and cloud development. When clouds are formed and developed as a result of the buoyancy of the air and thus atmospheric instability, vertical development takes place. Thus cumulus clouds are vertical clouds developing on a buoyant column of air. Stratus clouds are formed by the forced lifting of air, primarily in cyclonic storms, but the lifting can also be accomplished orographically.[3] This type of lifting develops sheets of clouds or stratus layers.

Precipitation

Precipitation is the descent through the air to the ground surface of water in some form. The qualification of reaching the ground is added, as some condensed moisture actually precipitates from cloud formations but does not reach the ground due to evaporation. This is most easily observed in arid or semi-arid regions where the condensation occurs at fairly high altitudes and the surface air has relatively low humidity.

It has already been pointed out that clouds are made up of condensed water vapor in very small particles. As long as the particles remain small so that suspension is maintained no precipitation will occur. There will be condensation at the cloud base and evaporation at the top and sides

---

[3]Orographic lifting is lifting of the air produced by some topographic barrier such as a coastline, range of hills, or mountains.

of the cloud formation such that there is a fairly constant cloud layer existing. When excessive condensation occurs and particle size increases by accretion to the point that the terminal fall velocity of the particle exceeds the upward velocity of the air then the water droplets will descend through the air column. A good sized raindrop may contain several million times as much water as an initial cloud particle. There is an upper limit to the size that a raindrop will grow. The maximum size depends on the velocity of the updraft near the ground and the size of the particle at the time it begins its descent. The largest drops often occur when large drops formed at the top of an updraft are rapidly shifted into a downdraft. Here friction is minimal and the droplet descends with only a small amount of disturbance.

There is some association between the type of precipitation which occurs and cloud form. Drizzle and light snow occur primarily from stratus forms while heavy snow, sleet, hail, and intense rain are often associated with cumulus cloud forms.

vertical transport of water vapor

In order for precipitation to occur there must be some mechanism present to transport large volumes of moist air upwards.[4] There are three primary mechanisms for lifting the air that account for most precipitation: convective lifting, cyclonic lifting, and orographic lifting. Convective precipitation is typical of warm moist air and is instigated by heating from the ground surface. When the surface is very warm the air immediately above is heated. It then expands and, having a lower density than the surrounding air, becomes buoyant. If the air becomes buoyant or unstable the lifting may continue until condensation and precipitation occur. Convective precipitation is highly variable in intensity, producing light showers as well as some of the heaviest downpours that occur in the atmosphere.

Cyclonic precipitation results from the lifting of moist air in a convergent wind system. In mid-latitudes this type of precipitation is considered to be the same as frontal precipitation since many if not most of the mid-latitude convergence systems have atmospheric fronts associated with them.

Orographic precipitation is either cyclonic or convectional in nature but it is intensified or increased by topographic lifting along coastlines, hills, and mountain ranges.

---

[4]An exception to this is some of the forms of "rainmaking" by man where the air is cooled *in situ*.

The total amount of precipitation which is received at the surface depends upon a number of factors. They are:

1. Whether convergence or divergence exists in the area.
2. Air mass origin, which is an indication of the temperature and moisture conditions of the air.
3. Topographic conditions.
4. Distance from the moisture source. The greater the distance from the source of the moisture the less water vapor will be present in the air due to prior precipitation loss.

Combining the above factors, the areas where the precipitation is greatest are mountain areas of the tropics where there is frequent convergence of air off the oceans. Where such conditions exist rainfall may reach 1020 cm (400 inches) a year and over.

Chapter 4

Atmospheric
Pressure and Wind

The Nature of Atmospheric Pressure

When a diver descends into water the pressure on him increases due to the weight of the water above. A cubic foot of water weighs approximately 62.4 pounds, so that for each foot a diver descends the weight of water above increases at a steady rate of 62.4 pounds per square foot of surface. The deeper he goes the greater the pressure. A person diving into a swimming pool may observe a feeling of pressure on his eardrums and on occasion even severe pain due to the sudden increase in pressure. The atmosphere also has mass so it exerts pressure much the same as water does. At the bottom of the atmosphere, the zone in which man lives, the pressure due to the mass of air lying above the surface averages about 14.7 pounds per square inch (1033 g/cm^2).

The atmosphere differs significantly from water in that it is far more compressible than water. The result of the compressibility is that the density of the atmosphere is far greater near the earth's surface than it is aloft due to the weight of the air above. The change in pressure with height is a geometrical rather than arithmetical factor. The rate of pressure with height is greatest near the surface and decreases at a decreasing rate with height. At low elevations the average decrease in pressure with height is equivalent to about 80 mm/km (one inch of mercury per thousand feet). As a result of the decreasing density with height, approximately one-half the mass of the atmosphere lies below 5500 meters (18,000 ft) and three-fourths lies below 10,000 meters (35,000 ft). Some atmospheric particles exist out hundreds of miles, however, in a gradually thinning zone. Changes in atmospheric pressure are usually not noticeable unless making a rapid descent or ascent through the atmosphere. This is understandable since the normal variation which occurs near the surface is restricted to something less than three per cent. Only in the presence of severe storms or exceptionally well-developed masses of cold air does the pressure vary more than the three per cent from the average, and even then ten per cent variation occurs only rarely in the lower atmosphere. Though the variation is relatively small, control of the spatial properties of the atmosphere are often a result of pressure variations. A small change in pressure may change the direction and velocity of the wind and subsequently lead to temperature and moisture changes. Air flow or wind which results from pressure differences serves two fundamental climatic functions: (1) to transfer heat from low to high latitudes, and (2) to transfer water from sea to land.

Mechanics of Horizontal Pressure Variations

Pressure changes within the atmosphere result from thermal and dynamic forces. Solar heating of the ground surface produces changes in air pressure. The heating of the air causes it to expand, becoming less dense and buoyant causing it to rise while more dense air flows in to replace it. Thermal currents produced by unequal heating develop at a variety of scales. Occasionally large land masses of subcontinental size are affected in this manner and such small units as plowed fields produce thermal currents.

The motion of the atmosphere itself is sufficient to produce pressure differences. Anyone who has observed a rapidly moving river has noticed the turbulence in it. Instead of all the water particles and other solid particles moving downstream in a smooth line there are horizontal counter currents as well as vertical movements. The same situation occurs within the atmosphere. Friction between the atmosphere and the earth, internal differences in the atmosphere, and the rotation of the earth combine to produce motion and pressure differences.

A pressure gradient is a change in pressure which occurs between two points or areas, generally in a horizontal plane. The pressure gradient represents the slope of the pressure surface much as the gradient of the land surface represents the slope of the land surface. A steep pressure gradient represents a rapid change in pressure over a short distance. A weak pressure gradient in turn is an indication of a minor change in pressure with distance. Weather maps show the pressure gradient on them by the use of isobars which are lines connecting points of equal barometric pressure. An area where the pressure gradient is steep will be shown by closely spaced isobars and areas of weak pressure gradients by widely spaced isobars.

Wind is a direct result of pressure differences. Unequal pressures in any fluid produce internal movement in order to redistribute the pressure equally. When a pressure gradient exists between two points there will be airflow between them. The airflow will be down the pressure gradient just as water will normally flow downhill along the steepest slope. The steeper the pressure gradient the higher the wind velocity is. If the pressure gradient were the only factor affecting the wind then the wind direction and velocity would indicate directly both the direction and steepness of the pressure gradient. When wind velocities are low and the pressure gradient weak the air flows directly down the gradient.

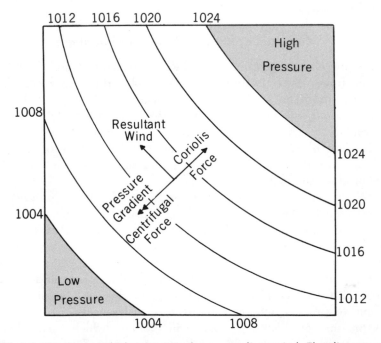

FIGURE 4.1 The forces which act to produce a gradient wind. The diagram represents a Northern-Hemisphere situation.

A second factor which affects wind direction and velocity is the rotation of the earth on its axis. The rotation of the earth imparts a rotational aspect to the wind which is referred to as the Coriolis force after G. G. Coriolis, the French scientist who first expressed it mathematically in 1844. This rotational aspect is imparted to all objects or media traveling free from the surface of the earth, and is an apparent motion rather than a real motion. It can be treated as if it is a force since it can be determined precisely for a given set of conditions. The amount of rotation of the wind is a function of the latitude and wind velocity. The Coriolis force increases with latitude and with wind velocity. The result of the Coriolis force is that in the northern hemisphere the wind appears to veer clockwise or to the right from its source, and in the southern hemisphere the rotation is counterclockwise or to the left.

A third force acting on the wind is the centrifugal force produced by movement of the air along a curved path. This force is exerted outward from the center of rotation. When the Coriolis force, pressure gradient, and centrifugal force are in balance the air will flow perpendicular to the pressure gradient and parallel to the isobars of a synoptic chart. This is defined as a gradient wind. When the isobars are straight and parallel, indicating a fairly uniform pressure gradient over a large area, the centrifugal force will be absent and the Coriolis force alone balances the pressure gradient. The resultant wind is a geostrophic wind and is found where wind velocities are high and the pressure gradient moderate to steep. Winds aloft far enough to be away from the frictional drag of the ground surface are usually geostrophic.

The fourth force affecting the direction and velocity of the wind is friction. Friction between the land or water surface and the air retards wind velocity and a reduction in wind velocity reduces the rotation produced by the Coriolis force. The pressure gradient becomes more significant in determining wind direction when the Coriolis effect is reduced. Most winds at the ground surface are a result of all three factors: the pressure gradient, the Coriolis force, and friction. As a result the winds blow down the pressure gradient at an angle which varies with the velocity. The lower the velocity of the wind the more directly the air will flow down the gradient. The frictional forces affect the wind to heights of about one kilometer above the surface under normal conditions.

Wind velocity increases with height in the lower troposphere. Near the ground surface itself the velocity is close to zero and the greatest increase in velocity takes place the first 30 to 60 meters (100 to 200 ft) above the ground. The velocity at 30 meters (100 ft) is usually about four times that at 50 cm (1½ ft) and is extremely gusty due to the obstructions on the ground. This surface turbulence decreases rapidly above the first few hundred meters but goes up to about 1000 m (3300 ft). The turbulence that exists at higher altitudes is due more to internal differences in the air than to surface friction. In the upper troposphere the winds are geostrophic and there are several very strong west to east currents. The most common of these jet streams are associated with the westerly winds of mid-latitudes. These easterly moving streams of air are in excess of 1000 meters thick and 500-850 km (300-400 mi) in width. These streams which lie at a height of some 8-13 km have velocities in the center ranging up to 500 km per hour (300 miles per hour). The northern hemisphere jet stream undulates north and south and extends furthest equatorward over the land masses and on some occasions the stream splits into two or more individual streams. The velocities are strongest in the winter when there is a much steeper energy gradient between the equator and the poles. Several other high altitude streams have been located in addition to those of the mid-latitudes.

Annual and diurnal variations in pressure and winds exist in many parts of the world. Wind velocities are greater in winter and spring due to the greater contrast between temperatures of high and low latitudes. Seasonal changes due to relative differences in land and water temperatures are also common. Regular diurnal variations in wind velocities are associated with changing ground temperatures through the day. The difference in heating in the daylight hours is greater than the difference produced by cooling at night. Thus diurnal winds tend to be much stronger in the daytime component than in the night component. These local winds can be obliterated at any given time by other and larger weather systems.

One of the winds which is associated with temperature differences of the ground surface and that is periodic in character is the land and sea breeze associated with coastal areas and shorelines of large lakes. The land and sea breeze is a function of the change in temperature and pressure of the air over land in contrast to that of water. In the hours after sunrise, land temperatures rise rapidly compared to the nearby water. With the heating of the overlying air above the shore-

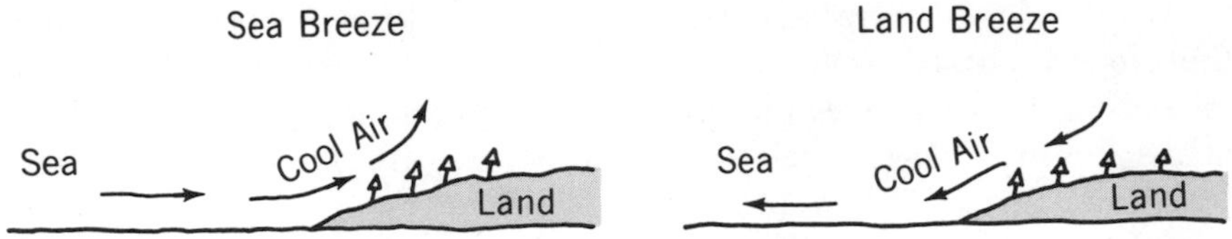

Sea Breeze   Land Breeze

FIGURE 4.2 Land and sea breeze

line the air expands and density decreases. The air rises over the land surface and air flows in from the water to replace it. This daytime flow of air landward is the sea breeze. The sea breeze usually begins several hours after sunrise and peaks in the afternoon when temperatures are highest. This coastal circulation will penetrate up to 50 km (30 mi) or more along the ocean and to lesser extent around lakes. The cell is quite shallow, ranging up to several kilometers (2 mi) deep under optimum conditions. The sea breeze is generally gusty and brings a decrease in temperature and an increase in humidity. Around lakes the same phenomena may occur but on a smaller scale. The complementary land breeze is less well-developed than the sea breeze. It occurs when the land surface cools at night and the water surface remains warm. The temperature and pressure relationships are reversed, but to a lesser degree, as the water surface does not actually gain heat at night but retains the heat acquired during the daylight period and so remains relatively warm with respect to the land.

Blizzards and dust storms of mid-latitudes are both associated with another mechanism. They are caused by rapidly moving masses of cold dense air from the polar regions. The blizzard is essentially blowing snow where the wind is produced by the turbulence along the edge of the cold air mass. These storms are often accompanied by sharp drops in temperature with the arrival of the cold air. Great Falls, Montana recorded a temperature drop of 30° C (55° F) in three hours as the result of the passage of one of these outbursts of cold air. Dust storms can be produced by any kind of wind of high velocity including those occurring with thunderstorms and tornadoes, but the extensive dust storms of the midwest result from the same conditions as the blizzard.

One of the products of differential heating and internal motion in the atmosphere is the development of regions characterized by above or below average pressure. The areas with below average pressure are also areas of convergence. Convergence exists when air currents are merging or moving on intersecting paths. This mergence brings air from different parts of the atmosphere together. As the air streams merge at the surface the only outlet for them is aloft. Areas of divergence occur when there is an outward flow of air and the air streams are separating. As the air spreads out laterally at the surface there is a replacement of air in the source region by subsidence from aloft. If there is an area of convergence at the surface there must necessarily be an area of divergence aloft, to maintain continuity of flow.

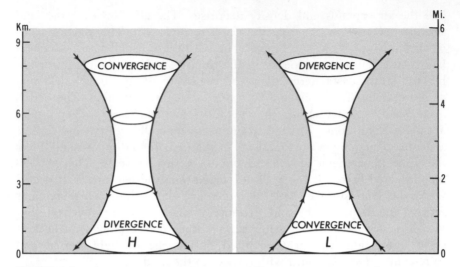

FIGURE 4.3. Convergence and divergence. When a strong system of convergence exists at the surface divergence is usually found aloft.

Between latitudes of about 40° N and 35° S incoming solar radiation exceeds earth radiation to space and in the latitudes poleward from these two points there is a net loss of energy to space. The primary difference in the radiation budget with latitude is in terms of incoming shortwave radiation. The equatorial region receives about 2.5 times the amount of solar energy that polar areas receive. Outgoing longwave radiation varies to a much lesser extent from equator to pole. In order for earth radiation to exceed solar radiation in polar regions, there must be an energy flux from tropical to polar areas. Without a latitudinal flux of energy there would be much higher temperatures at the equator and much colder temperatures at the poles than is presently the case. The poleward transfer of energy is by the atmosphere and the ocean with about 80 per cent of the total transfer carried out by the atmosphere. The primary circulation of the atmosphere is the mechanism by which the energy is carried poleward. This complex circulation system consists of a number of semi-permanent areas of convergence and divergence and the airflow in and between them. The energy is transported primarily as sensible heat and latent heat of water vapor. The kinetic energy contained in the circulating wind systems is eventually converted to sensible heat either by internal friction or by friction with the ground surface. Another example of equilibrium in the environment is present in this case as the rate of kinetic energy generated within the atmosphere is balanced by the energy lost to the system by friction.

If the earth were not rotating there would be a much simpler circulation of air but the earth is rotating and this rotation plays a very important part in determining the circulation as it exists. Rotating bodies are subject to the law of conservation of momentum. The law states that if no external force acts on a system the total momentum of the system is unchanged. Angular momentum is a function of the mass of the rotating body, the angular velocity (degrees per unit time) and the radius of curvature. The product of these elements tends to stay constant without external interference. A tetherball serves as a good illustration. As the ball goes around the pole and the radius shortens the angular velocity increases to keep the momentum the same as it was initially. Since air has mass, it also has momentum, As it moves poleward or equatorward, the rotational velocity changes as its distance to the earth's axis changes. As air moves poleward it is given an eastward velocity due to its shorter radius and hence becomes a westerly wind. In actual practice the increase in angular velocity is much less than it might be as the energy is dissipated by friction and by diffusion in the large mid-latitude eddies or cyclones.

George Hadley in 1735 proposed a cellular structure for the atmosphere which included the subsidence of cold air at the poles and an equatorward surface flow and with subsequent rising of the air at the equator as it is heated and then poleward flow aloft. The rotation of the earth prevents the development of such a single cell and Hadley's model was gradually altered to include a three cell structure between the poles and the equator, still with subsiding cold air at the poles and rising warm air at the equator. This three cell structure, still in the process of refinement, illustrates in a crude way the essentials of the primary circulation. Thus, the heat differential between the equator and the poles drives the general circulation and the earth's rotation determines the complex pattern of circulation that results from the energy differential.

The Tropical Circulation

The most equatorward of the three cells most resembles that proposed in the three cell model and also resembles the original circulation cell proposed by Hadley. The cell extends from the vicinity of the respective tropic to near the equator and features surface flow towards the equator and counterflow aloft with rising air at the equator and subsiding air near the tropics. This section of the general circulation is still referred to as the Hadley cell as it operates effectively as Hadley outlined it in 1735, even if in a more complex manner.

Near the equator the surface and the atmosphere above it are heated excessively by solar radiation through the year, to the extent that a low pressure zone results which is intensified by the circulation itself. The air flowing into the zone is of similar temperature and moisture characteristics. A weak pressure gradient produces light and variable winds. Where the converging air has had a trajectory over the ocean, it contains large amounts of moisture, and cloudiness and frequent precipitation are common to the area. This convergence zone is often referred to as the intertropical convergence zone, or ITC, as it represents the zone in which the trade winds from north and south of the equator merge. In the center of this zone are found the tropical easterlies, which are fairly stable winds of low velocities moving from east to west. These are not to be confused with the Trade Winds which are discussed later in this section. These easterly winds are quite regular in direction if not velocity. Contrary to popular belief the weather of the tropics is not monotonous but in fact quite varied.

Occasionally a change in circulation will bring quite unstable weather in the form of squalls and general rainstorms. The converging winds on occasion will rise some distance away from the center of the convergence zone. When this happens quite stagnant conditions may result near the surface. These stagnant conditions of very low velocity winds with ill-defined direction were named the Doldrums by sailors of the sixteenth century. Sailing vessels while traveling to the western hemisphere from Europe were occasionally becalmed in this area just as they were further poleward in the Horse Latitudes. At approximately 30° N and S of the Equator is a semi-permanent belt of surface divergence and above average sea-level pressure. This zone is a result more of the motion of the atmosphere and the Coriolis force than from thermal factors. The high pressure that prevails is due to the subsidence of air from upper levels, and the subsidence gives rise to divergence at the surface. The term Horse Latitudes found in historical literature applies to this area of divergence. The zone is similar in some respects to the equatorial convergence zone in that the pressure gradient is usually weak and resultant winds are light and variable in direction. The moisture characteristics of the zone are quite different from the equatorial zone as the subsiding air has little moisture in it. Further, it is more intense next to western continental margins: apparently earth rotation causes air to "pile up" against land masses and dryness is exaggerated in those regions.

Lying between the subtropical belt of divergence and the zone of equatorial convergence is a region of surface winds flowing equatorward known as the trades. This wind system is best developed in a belt some 10° in width, centered on 15° either side of the Equator, but the location shifts with the two adjacent zones. This wind belt represents one of the more regular wind systems found at the surface of the

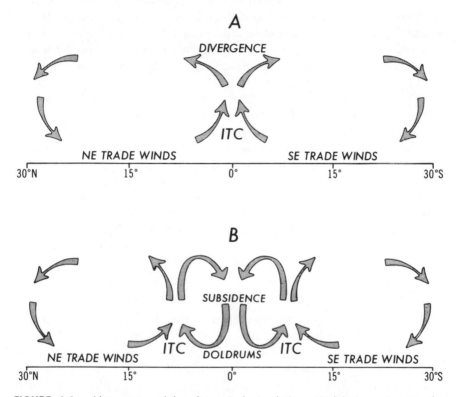

FIGURE 4.4. Alternate models of tropical circulation. Model A represents the tropical circulation as it is usually represented with a single Intertropical Convergence Zone. Model B illustrates the tropical circulation with a double ITC.. Both systems could occur at different times.

earth. These winds, which average from 4-7 kilometers (3-4 mi) per hour are best developed towards the eastern half of the oceans and are more dependable there than elsewhere. Both poleward and equatorward from 15 degrees these winds are less distinct. Their origin is in the subsidence of the subtropical highs and they are eventually lost in the convergence of the equatorial low. The Coriolis force gives the winds a westward component so that they are generally northeasterly in the northern hemisphere and southeasterly in the southern hemisphere.[1] The Trade Winds are often layered, with a surface layer that becomes increasingly moist and unstable as the air moves towards the equator and an overlying layer which is fairly dry and stable. As the air moves equatorward it not only increases its moisture and instability but adds latent heat at the same time. Cloudiness and precipitation increases with distance from the origin and the trades bring precipitation to the windward coasts in the affected latitudes. These winds are predominantly oceanic winds and they do not usually penetrate great distances inland. This zone, nevertheless, represents the major easterly wind of the troposphere and has its maximum component at 15 degrees and a height of one to three kilometers (2 mi).

Subsidence and hence stability is much greater over the eastern side of the subtropical high. The trades blowing out of the east side also have a shorter travel distance over the ocean so they are drier and more stable than those further west. It is the strong divergence and easterly winds along the east side of the subtropical highs which gives rise to the cold currents and upwelling of cold water associated with the coastlines in the area. Above the trades are the anti-trades, poleward flowing air concentrated at 10-12 kilometers (6 to 7 miles). These winds are extremely variable and represent a poleward flow only in terms of net motion. They increase in velocity as they go poleward and form a subtropical jet stream above the subtropical high pressure cells.

Mid-Latitude Circulation

As suggested in previous paragraphs, the three cell model of the atmosphere has inadequacies and its departure from the real world is greatest in mid-latitudes. The model calls for a mean poleward flow of air between the semi-permanent high pressure cells at 30° and the semi-permanent low pressure trough near 60°. In reality the mean meridional flow in these latitudes is very weak. In mid-latitudes surface winds are quite variable in both direction and velocity but the greatest frequency and highest average velocity indicates a west to east flow. Maximum westerly velocities are reached at about 35° North and velocities at this latitude becomes more pronounced with height up to the tropopause. Thus, the primary flow in mid-latitudes is zonal, i.e., west to east, but it is in the latitudes of 35°-40° North and South that the maximum poleward transfer of energy takes place. With a

---

[1]Winds are named for the direction *from* which they blow. A *northerly* wind blows from north to south.

relatively weak meridional flow there must be some other means of transporting the energy through the zone. Located in this latitudinal range is the transition zone between the warm tropical air and the cold polar air known as the planetary frontal zone. Here the mid-latitude westerlies are strongest and the flow increases with height up to about 12 km. In the upper part of the troposphere and lower stratosphere long waves develop in the planetary frontal zone that meander around the earth. Having an amplitude wavelength which varies up to about 6000 km., these waves are always in a state of flux, sometimes as standing waves and sometimes in motion with the westerlies.

Near the surface large eddies or vortices develop that are from 1000 to 2000 kilometers (600-1200 mi) across and have lifetimes of several days. These eddies are the cyclones and anticyclones of the mid-latitudes. Being rotational in character they carry cold air from the poles southward and warm air from the subtropics northward. The cold fronts and warm fronts that make up much of the center of attention on weather forecasts are the leading edges of the airflow being carried across the polar front. These eddies thus carry energy poleward while the net flow of air itself is very low. It is really an exchange process of warm for cold air. Thus, there is a high energy transfer with a low net poleward flow of air. It is also the high frequency of these cyclonic cells which forms the subpolar low pressure zone. This convergence system differs markedly from the equatorial convergence zone in terms of the kinds of air flowing into it. At this latitude the convergence is one of cool dry air from the arctic areas and warm moist air from the tropics. This is the reason for the marked changeability of mid-latitude weather. Because the changing winds associated with the traveling vortices are so different in temperature and moisture characteristics the areas in the latitude of the polar front have such drastic and frequent changes in weather.

Polar Circulation

Over the north and south polar regions thermally induced high pressure systems exist although knowledge regarding them is rather limited. The high pressure cells are usually displaced away from the geographic pole towards the continental cold pole. This displacement in the Antarctic is not large but in the northern hemisphere it is significant. In the northern hemisphere winter there is always a heat flow from the open leads and also through the pack ice so that the cold pole is shunted off on the land mass. Since there are two major land masses in the northern hemisphere there are often two cold poles, one over North America and one over Eurasia. Upper air divergence and temperature inversions are a characteristic. There is a predominance of high pressure systems and divergent air flow over polar areas most of the year but the size of the area affected changes, expanding during the winter months. Frequently found outward from the polar highs are belts of weak easterly winds that tend to be more pronounced in the northern hemisphere during the summer. They are stronger around the Antarctic continent being reinforced katabatically. Around the Antarctic continent and Greenland are strong gravity winds blowing down the ice slope and out to sea. They can attain fairly high velocities and be quite persistent but they are usually shallow, extending upwards to some 300-400 meters (1000-1300 ft).

The tropical Hadley cell and the polar cells to a lesser extent are driven by the heat imbalance between the tropical and polar areas. The middle latitude cell, such as it is, operates mechanically. Since the basic driving mechanism is solar energy it stands to reason that the shifting of location of maximum receipt of solar energy would affect the location of the general circulation. The differential rates of heating land and water surfaces also affects the location of the general circulation. The greatest share of the earth's land mass lies north of the equator and this tends to offset the whole general circulation northward from the equator. Here it is perhaps best to distinguish between the geographic equator at 0° latitude and the thermal equator poleward several degrees. The greater intensity of the antarctic land mass also plays a part in shifting the system northward. The earth surface is more uniform in the Southern Hemisphere and the temperature gradient is steeper and more uniform in the mid-latitude areas. Thus, the circulation system is better defined and stronger in the Southern Hemisphere. The wind systems, especially the westerlies, are of higher velocity in the southern than in the northern hemisphere. In the Southern Hemisphere belt of westerlies the winds are given such names as the roaring forties, furious fifties, and shrieking sixties. Winds tend to be more persistent over the oceans in both hemispheres than over the land areas. The extent to which the ITC migrates following the sun is essentially a function of the nature of the surface. Radiation received at the ocean surface is primarily used to evaporate water so the surface temperature does not change much nor does the air above it. On the continental land masses it is a different story. The rate of absorption is five or six times as great providing a considerable amount of energy for the lower atmosphere. The displacement of the ITC reaches its greatest extent over the Eurasian land mass where the migration reaches 30° N over India. Recent evidence suggests there may be two convergence zones, one on either side of the equator rather than a single cell migrating back and forth over the equator. If a double convergence zone exists, the intensity probably varies with the season, being more intense in the summer hemisphere.

The other wind and pressure belts shift also with the seasons giving rise to rainy and dry seasons as well as modification of the temperature regime.

Upper Air Circulation

The troposphere receives most of its energy from the ground surface in the form of longwave radiation and the heat of condensation. The upper atmosphere receives most of its energy from the direct absorption of solar or earth radiation. The lower stratosphere exhibits the basic circulation patterns of the upper troposphere, that is, the large storm systems penetrate the tropopause and dictate the flow pattern. In the latitudes astride the subtropical highs the mid-latitude and tropical tropopauses are separate and overlap. Between the two tropopauses is a latitudinal exchange of air. Tropical air from the troposphere is exchanged for polar air from the lower stratosphere. This exchange plus the air forced into the lower stratosphere by severe storms ventilates the lower stratosphere periodically. The upper atmosphere is characterized by zonal flow for the most part and in mid-latitudes the direction of flow tends to reverse itself with the seasons. Recent investigations have shown a biennial reversal of direction over the equator. Since data for high altitude winds are still sparse the explanation of some of these upper atmospheric winds is still not established. There are three warm zones and two cold zones between in a vertical section of the atmosphere. These alternating temperature levels provide a mechanism for vertical circulation. The lower one is well-established in the form of the tropospheric circulation. Knowledge of circulation in the cells above is limited but the mass of atmosphere involved is very small.

In addition to serving as a mover of heat the primary circulation is a mover of moisture. Moisture is transferred from low to high latitudes and from the ocean to the land masses. The oceans are the source for the moisture in the atmosphere and some parts of the ocean yield much more water to the atmosphere through evaporation than other parts. Evaporation will necessarily be greatest where there will be the

greatest differences between the vapor pressure of the air and the water. Ocean temperatures are the highest at latitudes approximately 30
degrees on either side of the equator. This is the result of high intensity
radiation and a clear atmosphere associated with the divergence existing here. As the air subsides from aloft it is quite dry being heated
adiabatically as it descends. There is a great difference between the
vapor pressure of the air and the water beneath due to the high temperature of the water and the low moisture content of the air. Evaporation is higher here than along the equator because there is more water present in the equatorial air and the vapor pressure at the water
surface is less. Sea temperatures are not as high along the equator because of the addition of fresh water at the surface from precipitation
and streams draining adjacent land masses. Insolation is reduced as
a result of the more extensive cloud cover which helps reduce water
temperatures.

In the regions poleward from 30 degrees both the vapor pressure
of the water and the vapor pressure of the air are reduced as temperatures drop and evaporation is consequently reduced. The subtropical
oceans in the vicinity of the divergence zones are the major source
regions for moisture which falls as precipitation on land areas of
low and mid-latitudes.

Associated with the primary circulation of the atmosphere are four
major zones of precipitation, or lack of precipitation, with intermedi-
ate transition zones. Near the equator is a zone experiencing convergence and precipitation in all seasons. The trade winds moving towards
the equator pick up moisture over the oceans and when lifted in the
equatorial convergence zone yield abundant precipitation. Precipitation
is increased over coastal areas by orographic lifting and increased convection started from surface heating. The average precipitation ranges
from 150-200 cm (60-80 inches) annually but in some areas exceeds
250 cm (100 inches). The higher amounts are generally found over
land areas and on windward coasts in particular.

The subtropical latitudes centered on 30 degrees are low rainfall
areas due to greater surface divergence. On the west sides of the continents at this latitude deserts exist due to the strong subsidence, and
annual rainfall totals drop below 25 cm (10 inches) a year. In some
locations the precipitation totals are so low that they are meaningless.
On the eastern sides of the land masses at the same latitude precipitation does not drop to such low amounts. Annual totals are more likely to be in the range of 50-75 cm (20-30 inches) or more. Poleward
from the subtropics precipitation increases, particularly on the west

sides of the continents, to a maximum in the subpolar convergence zone near 60 degrees latitude. Precipitation totals run above 150 cm (60 inches) per year in coastal areas. In mid-latitudes as in the tropics there is orographic intensification of precipitation. Windward sides of land masses are well watered, whereas the leeward locations have lesser amounts of precipitation. All three major precipitation mechanisms, convectional, cyclonic, and orographic lifting, produce precipitation. In the continental interiors distance from the ocean becomes a factor in determining the total amount of precipitation. This is due to the decrease in moisture content of the air by precipitation as it travels inland.

High latitude locations receive generally low precipitation amounts. Twenty-five to forty centimeters (10-15 in) is perhaps the most typical except in island or coastal areas with onshore winds.

Four basically different temporal precipitation regimes can be identified. In the two major convergence zones areas exist where precipitation occurs distributed uniformly year round. Associated with the centers of divergence is a regime of low frequency of precipitation year round. Most of the earth's surface is subject to a seasonal pattern and most often the maximum occurs in the summer, or high sun season. The fourth type of regime is the seasonal rainfall climate with the maximum being in the winter months. Areas experiencing this type of rainfall pattern exist in both the tropics and mid-latitudes.

The Effects of Land Masses on Pressure and Wind

The differing reaction of land and water to solar radiation has a marked effect on pressure belts in mid-latitudes. Those most affected are the subtropical high and the subpolar low. Seasonal reversals occur over the land masses in terms of pressure systems and resultant convergence and divergence. During the winter half year the subtropical high and the polar high expand and often meet over the land masses that have undergone substantial cooling. In mid-winter high pressure ridges or cells are often found over the northern hemisphere land masses. During the summer months the subpolar low expands over the continental areas and sometimes tropical lows and the subpolar low merge to form a single trough of low pressure extending from the tropics to 60 or 70 degrees of latitude.

The change in pressure systems brings about a marked reversal in wind and moisture conditions with the season. During the summer months convergence predominates and high humidity and precipitation are general over the continents. In the winter months when high pressure systems predominate, divergence over the land is more frequent than convergence. As a result of the latter humidity and precipitation are both lower in winter than in summer. This seasonal shift of pressure with all its ramifications is a monsoon.

The monsoon applies to North America as well as to Asia although it is most commonly associated with the latter. In summer the subtropical high shifts north over the Pacific coast and virtual desert conditions are maintained over much of southern California. The summer season is the dry season for all of United States west of the Front Ranges of the Rockies. On the eastern side of the continent the low pressure trough is more persistent and convergence is frequent over this part of the continent. Onshore winds from the Atlantic high pressure system centered over the Azores Islands carry considerable moisture. The convergence and subsequent lifting over the land areas produces an appreciable amount of rainfall and a summer maximum over much of the eastern part of the country.

In the winter the subtropical high over the Pacific moves equatorward as the vertical radiation from the sun moves southward and the sub-polar low moves southward as the subtropical high does. As the sub-polar zone of convergence moves southward precipitation increases along the west coast. To the east of the mountains the winter high pressure systems are more frequent and the Great Plains experiences a season of less precipitation although not absolute drought.

The seasonal shift in pressure, wind, and humidity over Asia is greater than it is over North America. During the winter dry subsiding air covers much of the continent and a dry season coincides. This dry season is comparatively much drier than the season is in North America. The subsidence produces offshore winds along much of coastal Asia. As these winds move out over the oceans they evaporate large amounts of water from the sea surface and serve as sources of moisture for the offshore islands. During the summer months the low pressure system and convergence is strong and the onshore flow of moisture is great. Moisture laden winds blow north from the Indian Ocean and the Arabian Sea. They converge over the northern plains and slopes of the Himalaya Mountains. One of the rainiest regions of the world is found here. Cherripunje India is often used as an extreme example of seasonal rainfall. Resulting precipitation on the coastal areas and oceanic sides of offshore islands is heavy. In the spring and fall there are intervening periods between the rainy season and the dry season when the weather is unsteady but not too unpleasant. The highest tempera-

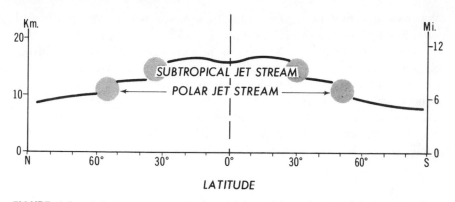

FIGURE 4.5. Jet streams occur in the vicinity of the subtropical highs as well as along the polar front. The tropopause is broken in the vicinity of the jet streams and occurs as a series of overlapping layers rather than as a continuous single layer.

tures of the year often occur late in the spring or in early summer before the high rates of moisture influx begin. The clear skies prior to the development of the onshore winds allow more insolation than does the maritime air which follows later. The rains and humidity which reduce absolute temperatures actually increase sensible temperatures.

The monsoon is probably the product of a variety of factors working together. The ITC migrates to between 25° and 30°; the land mass interior records a large change in pressure that enforces the migration of the ITC; and it is likely that Himalaya Mountains and the mid-latitude jet stream play a role. The Himalaya Mountains is a very significant topographic barrier and tends to split the circulation over Asia. In the winter the jet stream is most frequently located over the mountains. At this time the polar high is expanded and covers much of north-central Asia. Divergence from this strong system north of the Himalayas produces extremely strong offshore surface winds. South of the Himalayas and the jet stream the anti-cyclonic flow is much weaker though offshore flow prevails. In summer as the temperature differential over the continent diminishes the jet stream breaks down and the subtropical high and ITC shift rapidly northward. This brings the sudden shift of winds onshore over the Indian Subcontinent. At the same time north of the Himalayas a weaker convergent system develops with attendant onshore flow of air. The precipitation is primarily convectional rainfall which is essentially distributed randomly within the flowing moist air. Some traveling cyclones associated with the ITC add further rainfall. The peak season of rainfall moves northward over the subcontinent as the zone of convergence moves northward towards the slopes of the Himalayas.

The largest share of the rainfall in northern India occurs during the summer onshore monsoon. The Asian monsoon is quite variable from year to year. It is variable in terms of the time of year when the onshore and offshore flow begins, as is the duration and intensity of the rainy season. The total amount of rainfall varies as a consequence and the agricultural economy of Northern India and Pakistan are cruelly subject to the whims of the system.

Air Masses

The movement of air in the primary circulation is of quite different characteristics in different parts of the atmosphere. These differences which are most noticeable in terms of moisture and temperature have led to the development of the concept of air masses. Air masses are large bodies of air of considerable depth which are relatively homogeneous horizontally in terms of temperature and humidity. They form over large areas of water or land which are fairly uniform in temperature. The air over these surfaces assumes the temperature and moisture characteristics of the underlying surface, or source region. The primary circulation moves the air away from the source region and may move the air several thousand miles.

Two basic categories of air masses are recognized on the basis of temperature and two based on moisture. Air masses are classified as tropical if their source is in low latitudes and polar if the source region is in high latitudes. Air masses originating over land masses, therefore being relatively dry, are labelled *continental air masses;* those originating over oceans, being moist, are labelled *maritime air masses.* The result of this categorization is the identification of four individual kinds of air masses: maritime tropical (mT), maritime polar (mP), continental tropical (cT), and continental polar (cP).

The maritime tropical air masses are most frequently found on the margins of the subtropical divergence zones over the oceans; the continental tropical air masses originate in the subtropical divergence zone over land masses; the polar maritime air masses are associated with the convergence over high latitude oceans; and the polar continental air masses with the divergence of the polar land masses. Two additional categories are used in reference to the extremes of the continental polar and maritime tropical air masses. Continental Arctic (cA) air is used to indicate exceptionally cold dry air and Equatorial (E) for very warm moist air.

As soon as an air mass begins to move modification of the temperature and moisture characteristics begins. The air masses are modified in a variety of ways. They are altered by heating or cooling from the surface, either by conduction or radiation. Moisture content is altered by addition or removal of water by evaporation, condensation, and precipitation. Major modification is also produced by large and small scale turbulence in the atmosphere. At any given time an air mass is a result of the nature of the source region and the changes which the air mass has undergone while in motion over the surface.

Chapter 5

Classification of
Climates

The atmosphere is extremely heterogenous in its major elements of temperature, moisture, pressure and wind. For example, in northern Africa a temperature of 57.8° C (136° F) has been recorded while near the South Pole on the Antarctic continent a low temperature of −87.2° C (−125° F) has been recorded. Large variations in the amount of moisture present in the atmosphere are indicated by the wide range of precipitation totals that are recorded. Parts of the Atacama Desert of South America have never recorded a measurable amount of rainfall while other parts of South America, and other areas also, record more than 1000 cm (400 inches) each year as an average, and in some years the amounts have approached 2540 cm (1000 inches).

In view of the differences which exist from place to place in temperature, precipitation, and winds, it is unlikely that two places on the surface of the earth will have exactly the same kind of weather. Since there are an infinite, or unlimited, number of places on the earth's surface, it can readily be seen that the number of different types of climate that exists is very large. It is, in fact, impossible in the study of climate to examine the combinations of weather that occur at every point on the surface of the earth. It is desirable and essential to devise some means of categorizing or classifying those places that are similar to each other in order to reduce the number of different areas, or climates, for purposes of study.

To produce a classification of any data that is to be at all useful it is necessary first to group those items together which present the greatest number of common characteristics, and then subdivide these groups on a similar basis until a satisfactory degree of subdivision is reached. In most areas of science attempts have been made to produce a classification based upon the most fundamental characteristics possible, rather than on elements which might be more easily observed but of less intrinsic importance. Sometimes a group of items doesn't lend itself easily to grouping or classifying. Attempting to group places on the earth with similar weather is much like trying to group students on the basis of height or weight. There are no clearcut boundaries to be found between the groups.

The major elements of the atmosphere exist at all places just as all students have weight and height. These elements do not distinguish a region by their presence or absence but by the difference in the amount of the element. For the air to reach a temperature of 57.8° C it must move upward through lower values as heat is added to the atmosphere. By the same token the air does not suddenly become 87° below zero C. The temperature must drop progressively to that point. As a result of the continuous nature of the temperature scale it can be discerned that all temperature values between the extremes, and the corresponding amounts of heat, have existed in the atmosphere at some time and place. There are certain times during each year when nearly all temperatures between these extremes exist at the same moment at some place or another. The same situation exists with the other elements. If records were searched carefully and if there were enough observation stations every possible amount of annual rainfall up to some 1000 cm a year could be found.

Spatial heterogeneity is a basic assumption in regional climatology, but the differences which occur from place to place are not without order. There is a rather systematic distribution of the weather elements. Since change with time and place is an integral part of the weather of the earth, classification of climate should be based on variability of atmospheric conditions rather than on the average conditions of the atmosphere. Some of the changes which take place in the atmosphere at a place are regular changes that occur at fairly fixed intervals of time. These periodic changes which are most commonly regarded or observed are the seasonal and daily changes. Of the two types, the seasonal changes tend to have a greater variation in type and amount than the diurnal changes, but there are exceptions.

Temperature and moisture are perhaps the most important elements of weather and climate and the introduction to climatic classification is based upon seasonal temperature and seasonal moisture regimes. The classification process can be initiated by dividing the atmosphere into three temperature regions, the tropical, mid-latitude, and polar. These three regions are defined in terms of air mass origin.

The tropical region is that area of the atmosphere where the air masses are predominately of tropical origin. The polar region is the area dominated by polar air masses and the mid-latitude region is that area in which tropical and polar air masses both play a major part in the weather systems. The boundary between the tropical and mid-latitude climates is represented by the equatorward limit of the movement of polar air masses and the boundary between the mid-latitude and polar areas is located at the northerly most point of penetration of tropical air masses. Since air masses are fickle in terms of the paths they take, the boundaries must be recognized as fluctuating lines or zones.

Moisture is the second important element of climate which can be used as a basis for subdivision of the atmosphere. A three-category subdivision can be made on much the same basis as with temperature. The three categories being the rainy regions, seasonal rainfall regions, and the dry regions. In terms of air masses the rainy regions are those dominated by maritime air masses year round, the seasonal rainfall climates are those experiencing marked seasonal changes from maritime to continental air masses, and the dry climates are those dominated by continental air masses year round. This trisection of the two variables produces a nine-category classification.

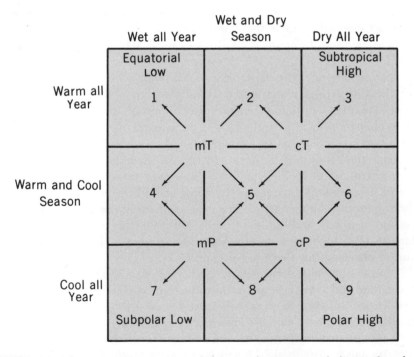

FIGURE 5.1 Schematic representation of the nine basic types of climate found on the surface of the earth. The seasonal patterns of temperature and moisture, the types of air masses most common to the area, and the location in the general circulation is shown.

Figure 5.1 shows the nine types established in block form to illustrate the relationship between some of the types. For example, each row and column of three has the same seasonal characteristic of either temperature or moisture. Thus for each of the three types of tropical climates there is a corresponding mid-latitude and polar climate with a similar seasonal moisture pattern, and so on.

Figure 5.2 illustrates several characteristics of four of the climatic types. These four climatic types tend to lack any strong seasonal variation in either temperature or moisture. Some differences between seasons exist, but they are not major. Each one of these four climatic types is controlled primarily by one kind of air mass, and each is to be found associated with one of the four semi-permanent pressure zones.

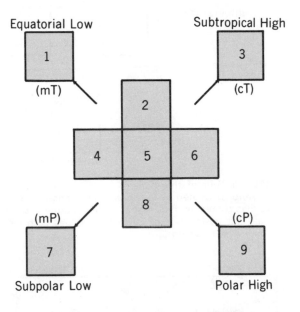

FIGURE 5.2 Four of the climates are controlled primarily by single types of air masses. Each of these is found in the core area of one of the semi-permanent pressure zones.

Four of the remaining five types of climate have a marked seasonal variation in either temperature or moisture, but not both. Two have pronounced seasonal moisture regimes and two pronounced seasonal temperature regimes. These four climatic types are situated between the major pressure zones and affected by the seasonal migration in these systems. They are exposed to air masses which are considerably different from one season to the next.

One of the nine climatic types, the mid-latitude seasonal rainfall climate, represents the maximum in seasonal variation of weather systems. Periodic invasions of air masses developing in each of the major source regions bring a great variety of weather.

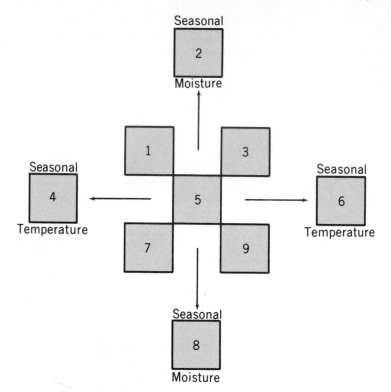

FIGURE 5.3 Four of the climates have a marked seasonal change in temperature or moisture associated with a seasonal change in air mass control. The other climate, that shown in the center of the diagram is subject to weather associated with all four basic kinds of air masses. It has both wet and dry seasons and summer and winter seasons.

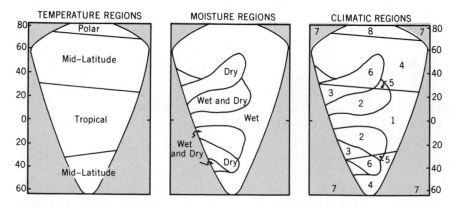

FIGURE 5.4 The distribution of temperature regimes and moisture regimes on a hypothetical continent. The combination produces the distribution of climates.

This system has four attributes: (1) it is based upon characteristics of the atmosphere itself; (2) it is based upon the dynamic character of climate rather than static conditions; (3) the boundaries are not emphasized; and (4) it is simple in construction.

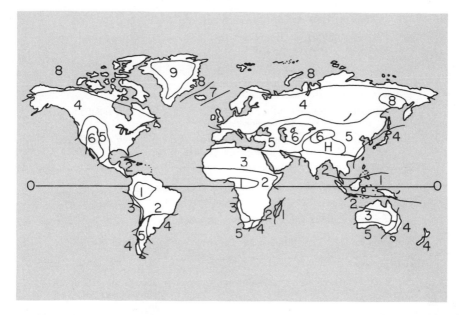

FIGURE 5.5 Actual distribution of the climates on the land masses of the earth. The distribution is highly generalized and the boundaries should be taken as approximate.

No major effort was made to classify climates from the time of the Greek civilization to around the start of the twentieth century, although there were during this interval numerous scattered efforts at collecting data pertaining to the atmosphere. Perhaps the best known system of climatic classification at the present time is that by Wladimir Köppen of Austria (1846-1940). His classification is based essentially on the distribution of vegetation. Köppen's assumption was that the type of vegetation found in an area is very closely related to the temperature and moisture characteristics of the area. These general relationships were already known at the time Köppen's classification was produced, but he attempted to translate the boundaries of selected plant types into climatic equivalents. The Köppen system is based on monthly mean temperatures, monthly mean precipitation, average annual precipitation, and mean annual temperature.

Table 5.1

The Köppen Classification

A Temperature of the coolest month above 18° C. (64.4° F.)

  Subcategories:

   f: rainfall in driest month at least 6 cm. (2.4 in.).

   m: rainfall in driest month greater than $10 - r/25$, but less than 6 cm. when r = annual rainfall in cm.

                 OR

      rainfall in driest month greater than $3.94 - r/25$, but less than 2.4 in., when r = annual rainfall in inches.

   w: rainfall in driest month less than 6 cm. (2.4 in.), but insufficient for m, and dry season in low sun period.

   s: rainfall in driest month less than 6 cm. (2.4 in.), but insufficient for m, and dry season in high sun period.

   w': maximum rainfall in autumn.

   w": two rainfall maxima, with intervening dry periods.

   i: annual temperature range less than 5° C. (9° F.).

   g: warmest month precedes summer solstice.

B Evaporation exceeds precipitation for the year.

  Subcategories:

   BS (Steppe): Derived by the following, when r = annual rainfall in cm. and t = annual average temp. in °C.

      70% of rainfall in summer six months: $r = 2(t + 14)$
      70% of rainfall in winter six months: $r = 2t$
      Even rainfall distribution or
          neither of above: $r = 2(t + 7)$

                     OR

                              when r = annual rainfall in inches, and t = annual average temp. in °F.

      70% of rainfall in summer six months: $r = .44t - 3.5$
      70% of rainfall in winter six months: $r = .44t - 14$
      Even rainfall distribution or
          neither of above: $r = .44t - 8.5$

      The value r is the BS/Humid boundary. When the derived r is greater than the value on the right of the equation the climate is humid; when less it is B. If B, then determine if BS by dividing the answer by 2. If, after dividing, r is greater than value on right, climate is BS; if less, climate is BW (desert).

   BW (Desert): Derived, as indicated above.

      h: average annual temperature above 18° C. (64.4° F.).

      k: average annual temperature below 18° C. (64.4° F.).

TABLE 5.1

(Continued)

k': average of warmest month below 18° C. (64. 4° F.).

n: high frequency of fog.

s: 70% of rainfall in winter six months (summer dry season).

w: 70% of rainfall in summer six months (winter dry season).

C Coolest month temperature averages below 18° C. (64. 4° F.) and above -3° C. (26.6° F.); warmest month is above 10° C. (50° F.).

Subcategories:

f: at least 3 cm. (1. 2 in.) of precipitation in each month; or, neither w or s.

w: minimum of 10 times as much precipitation in a summer month as in driest winter month.

s: minimum of 3 times as much precipitation in a winter month as in driest summer month, and one month with less than 3 cm. (1. 2 in.) of precipitation.

x: rainfall maximum in late spring or early summer; dry in late summer.

n: high frequency of fog.

a: warmest month over 22° C. (71. 6° F.).

b: warmest month under 22° C., but at least four months over 10° C. (50° F.).

c: only one to three months above 10° C.

i: mean annual temperature range less than 5° C. (9° F.).

g: warmest month precedes summer solstice.

t': hottest month delayed until autumn.

s': maximum rainfall in autumn.

D Coolest month temperature averages below -3° C. (26.6° F.) and warmest month over 10° C.

Subcategories:

d: coldest month below -38° C. (-36. 4° F.).

Other subcategories same as for C.

E Warmest month temperature averages less than 10° C. (50° F.).

Subcategories:

ET: average temperature of warmest month between 0° C. (32° F.) and 10° C. (50° F.).

EF: average temperature of warmest month below 0° C. (32° F.).

Köppen recognized four major temperature regimes, one tropical, two mid-latitude, and one polar. After identifying the four regimes he assigned numerical values to the boundaries. The tropical climate was delimited by a cool month temperature average of at least 18° C (64.4° F). This temperature was selected as it approximates the poleward limit of certain tropical plants. The two mid-latitude climates are distinguished on the basis of the mean temperature of the coolest month. If the mean temperature of the coolest month is below −3° C (26.6° F) it is microthermal, and if the temperature is above −3° C the climate is mesothermal. The fourth major temperature category is the polar climate. The boundary between the microthermal and polar climates was set at 10° C (50° F) for the average of the warmest month, which roughly corresponds to the northern limit of tree growth.

TABLE 5.2

Köppen's Major Climates

A Tropical Rainy Climates

B Dry Climates

C Mid-latitude Rainy Climates, Mild Winter

D Mid-latitude Rainy Climates, Cold Winter

E Polar Climates

Principal Climatic Types According to Köppen's Classification	
Af Tropical Rainy	Cw Mid-latitude Wet-and-Dry, Mild Winter
Aw Tropical Wet-and-Dry	Cf Mid-latitude Rainy, Mild Winter
Am Tropical Monsoon	Dw Mid-latutide Wet-and-Dry, Cold Winter
BS Steppe	Df Mid-latitude Rainy, Cold Winter
BW Desert	ET Tundra
Cs Mediterranean	EF Ice Cap

The system has been subjected to criticism from two aspects. There is no complete agreement between the distribution of natural vegetation and climate. This is to be expected since factors other than average climatic conditions (soil, for instance) affect the distribution of vegetation. The system is also criticized on the basis of the rigidity with which the boundaries are fixed. Temperatures at any site differ from year to year as does rainfall, and the boundary based on a given value of temperature will change location from year to year. In spite of the criticisms and the emperical basis of the classification it has proven quite usable as a general system.

C. W. Thornthwaite (1899-1963) published a classification of climate in 1931 which was based on the relationship of vegetation to climate, as was that of Köppen, but which was more systematic in construction. The two major ingredients in the classification by Thornthwaite are the concepts of precipitation effectiveness and temperature efficiency. Precipitation effectiveness is designed as an indicator of net moisture supply, taking into account both the actual amount of precipitation and the estimated consumption of moisture by evaporation. The precipitation effectiveness is determined by calculating the ratio of the precipitation to evaporation (P/E ratio) for each month of the year and summing them to form the precipitation effectiveness index (P-E index). Temperature efficiency in this classification is used as an indicator of the energy or

TABLE 5.3

The 1931 Thornthwaite Classification of Climate

Humidity Province	Vegetation	P-E Index
A Wet	Rain forest	above 128
B Humid	Forest	64-127
C Subhumid	Grassland	32-63
D Semiarid	Steppe	16-31
E Arid	Desert	under 16

Temperature Province	T-E Index
A' Tropical	above 128
B' Mesothermal	64-127
C' Microthermal	32-63
D' Taiga	16-31
E' Tundra	1-15
F' Ice Cap	0

Subprovinces of the Major Humidity Provinces
r rainfall abundant in all seasons
s rainfall deficient in summer
w rainfall deficient in winter
d rainfall deficient in all seasons

By Permission from "The Climates of North America According to a New Classification." *Geographical* Review, Vol. 21, October, 1931.

heat supply relative to evaporation rates. The index (T-E index) is cal-
culated in the same fashion as the P-E index using temperature and
evaporation data.

On the basis of the precipitation effectiveness index five moisture
provinces were established. The boundary of each progressively more
humid region is established at a doubling of the P-E index. On a comparable scale of the temperature efficiency index six major temperature
provinces are recognized. As in the case of the P-E index each progressively warmer province is bounded by an index double that of the
preceding province. Thornthwaite utilized many of the same letters that
Köppen used in his classification. To the two indexes of moisture and
temperature are added a letter designation for rainfall distribution
through the year. The initial classification yields some 32 different climatic types.

Each of the classifications is a means of extracting the systematic
pattern of climates on the earth's surface. The classifications have been
devised to reduce the great variety of climates that exist on the earth

EQUIVALENT KÖPPEN AND THORNTHWAITE CLIMATIC TYPES

	Wet	Wet and Dry	Dry	
Tropical	Af	Aw Am BSh	BWh	Köppen
	AA'r BA'r CA'r	BA'w CA'w DA'w	CA'd DA'd EA'd	Thornthwaite
Mid–latitude	Cf Df	Cs Dw Cw BSk	BWk	Köppen
	AB'r BC'r AC'r CB'r BB'r CC'r	BB's CB's DB's BB'w CB'wDA'w BC's CC's DB'w	CB'd DC'd CC'd EB'd DB'd EC'd	Thornthwaite
Polar		ET	EF	Köppen
		E	F	Thornthwaite

FIGURE 5.6 For each of the nine types of climates the categories in the Köppen
and Thornthwaite systems which are used to represent the climate are indicated.

into a manageable number of types. The climatic types that are identified do not stand as completely distinguished regions, but merge one into the other. No two areas are exactly alike in all their characteristics, nor need they be for the classification to be meaningful. Nor does the weather need to maintain exactly the same pattern year in and year out, for this is generally not the case. There are problems that result from placing a static classification on a dynamic atmosphere but there is enough regularity in the atmospheric circulation for the classifications to be usable for the purposes designed.

Chapter 6

The Tropical Climates

The primary circulation is responsible for the basic pattern of weather types which exist in the tropics. This circulation is essentially one of subsidence and divergence near the Tropics of Cancer and Capricorn and convergence near the Equator. The surface air flow is primarily of warm air, which lacks strong contrasts even when convergence occurs. Due to the extensive heating of equatorial regions, rising air tends to predominate over a fairly broad area near the Equator. Surface winds are of generally lower velocity, if not more predictable in direction, than in mid-latitudes.

The air which subsides in the vicinity of the subtropics is of fairly low relative and absolute humidity. If the air moves over the ocean, moisture is added to it by evaporation and the air develops a greater and greater tendency towards instability, with convectional clouds and precipitation being characteristic of the tropical maritime air.

The areas of divergence, convergence, and the intervening Trades shift latitudinally through the year following the thermal equator. This migration is responsible for the seasonal shifts in the moisture patterns of much of the tropics. The seasonal shift in moisture patterns is usually accompanied by a change in temperature.

Three major types of storms make up the secondary circulation in the tropics. If the primary circulation determines the regional pattern of atmospheric circulation and the seasonal regime, these storms account for most of the day-to-day weather. The storm types most widespread in the tropics are the thunderstorm, easterly wave, and the hurricane.

Thunderstorms are defined in terms of their electrical manifestations which in fact are a product of the storm itself. The thunderstorm is a convective cell of rather short life, developing in the atmosphere and accompanied by large vertical cumulo-form clouds, rain, lightning, and thunder.

The development of a thunderstorm is a continuous process but there are several recognizable stages through which the storm goes. The first stage is the cumulus stage. All thunderstorms begin as cumulus clouds, but not all cumulus clouds become thunderstorms. A cumulus cloud consisting of condensed moisture in the upper reaches of an updraft, or rising mass of air, is the first stage. Each of these cumulus clouds usually lasts but a few minutes and reaches heights of a few thousand feet. During this stage of growth there is a general updraft throughout the cloud formation and condensation in the form of water particles. The updrafts

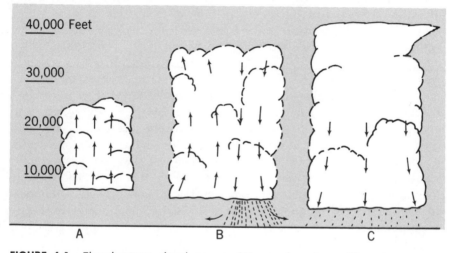

FIGURE 6.1 Thunderstorm development. (A) cumulus stage, (B) mature stage, (C) dissipating stage.

are strongest toward the top of the cloud, where velocities may be as high as 100 feet per second (over 60 mph). Temperatures within the cloud are usually higher for a given altitude than in the surrounding air.

The second stage occurs only if instability results at heights reached by the rising air column. The most intense storm activity exists in this stage. At this point rapid condensation takes place and precipitation begins within the cloud and subsequently may reach the ground. Precipitation may take place as high as 25,000 feet in a strong cell. The strongest updrafts and downdrafts are associated with the second or mature stage of the thunderstorm. It is during this stage that the rapid cauliflower growth takes place which may, in extreme cases, extend upward to the tropopause. Downdrafts are started by the friction of the falling precipitation and after they once start are self-perpetuating. Temperatures in the downdraft are cooler than those in the updrafts and the subsiding air spreads out along the ground around the storm due to its greater density. This is also the stage in the storm's history when lightning is most active. The greatest share of the lightning occurs at altitudes of 15,000 to 25,000 feet, and much is not visible from the ground unless the observer is at a considerable distance from the cloud.

The last major stage in the life cycle of the thunderstorm begins when the downdraft spreads over the entire cloud. This marks the dissipating stage because there is no longer moisture being carried aloft. The downdraft heats adiabatically and thus is unsaturated when it reaches lower altitudes. If several of these cells exist near each other new cells may be produced. The outward flowing cold air from two or more cells may collide and be forced aloft starting a new cell. This occurs often if two cells are within 3 miles of each other. Most thunderstorms consist of numerous individual cells forming and dissipating.

The surface weather associated with thunderstorms is characterized by intensive precipitation, decreasing temperatures, gusty winds, and changing wind direction. The rainfall first covers a very small area and then spreads out in all directions. The cold downdrafts can spread outward without carrying moisture with them, as the precipitation falls nearly vertically and is not carried outward by the wind. The average length of time that it rains at a point on the ground beneath a thunderstorm is about 25 minutes. The most intensive rainfall occurs shortly after the rain begins, usually within two or three minutes, and then begins to slacken gradually.

Dry thunderstorms occur in arid and semi-arid regions. These are called dry thunderstorms as there is not enough precipitation from them to reach the ground. In these storms the cool, dry, descending air sometimes reaches the ground and sometimes not. This type of storm is a hazard in the areas in which it occurs as it is often accompanied by cloud-to-ground lightning, which causes forest fires yet yields no rain with which to put them out.

Hailstorms are a special manifestation of thunderstorms. Hail forms when there is fusion or condensation into ice particles and where the particles are suspended for long periods. There are different theories on the formation of hailstones. However, it is generally agreed that a hailstone starts from a small rounded nucleus of either compacted, partially melted snow, or from a large raindrop in which freezing has already begun 1 or 2 miles above the ground. The freezing formation can increase to .25 inch in diameter as water droplets continue to freeze upon it.

A large storm system, if not a particularly violent one, which is characteristic of the tropics is the easterly wave. The easterly wave may be the tropical counterpart of the mid-latitude cyclone but there are several major differences in the two systems. The tropical wave lacks the sharp contrast in air masses associated with mid-latitude storms, and the Coriolis force responsible for producing curvature of the airflow is not very great.

Divergence and dissipating cumulus cloud forms mark the forward edge of the wave. Behind the wave convergence exists, with associated high cumulus forms, and the thunderstorms are often arranged as cloud streets. Cloud streets are long parallel lines of clouds that form with light winds of constant direction. The direction of airflow and associated cloud streets changes about 90° across the crest of the wave. The wave itself has only a slight pressure drop, something in the vicinity of two to three millibars. These waves travel rather slowly, usually less than 15 miles per hour. There is an extensive area of discontinuous precipita-

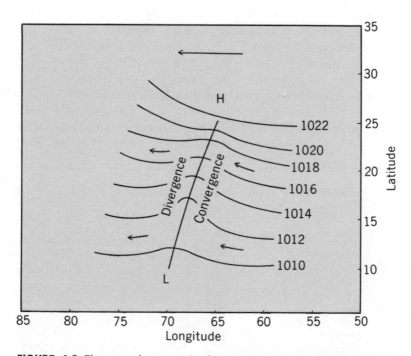

FIGURE 6.2 The easterly wave. A plan view of an easterly wave in the tropics showing the wave in the pressure, the areas of convergence and divergence and direction of storm movement and air flow through the storm. From **Tropical Meteorology** by Herbert Riehl, Copyright 1954, McGraw-Hill. Used by permission of McGraw-Hill Book Company.

tion east of the wave crest. The prolonged periods of drizzle and steady rain of the tropics are largely a result of these weak waves in the atmosphere. They are associated with the Trades, but sometimes they are found in latitudes between 20° and 30° either side of the Equator.

The third type of storm characteristic of the tropics is the tropical cyclone. The tropical cyclone is the typhoon of the Pacific and hurricane of the Atlantic. This storm is a large vortex or circular storm. It rotates according to the Coriolis force—counterclockwise in the Northern Hemisphere and clockwise in the Southern. These storms derive their energy from the latent heat of condensation. The energy input in an average hurricane may be equivalent to more than 10,000 atomic bombs the size of the Nagasaki bomb. These storms range in size from a few miles in diameter up to several hundred miles in diameter, with an eye in the middle which is up to 40 miles across. The total area involved may be as much as 200,000 square miles.

The atmospheric pressure tends to be symmetrically distributed about the center. The pressure may go as low as 880 mb., but this is rare. This represents a drop of about 13 per cent, which is not an explosive drop since it takes place over a distance of many miles.

The wind system associated with the hurricane is one of contrasts. In the eye of the storm the winds will be light and variable and of velocities not usually exceeding 15 miles per hour. The wind velocities increase rapidly away from the eye, reaching their highest velocities just outside the eye and at a height of about 1/2 mile. To be classified as a hurricane the winds must exceed 75 miles per hour. In most well-developed hurri-

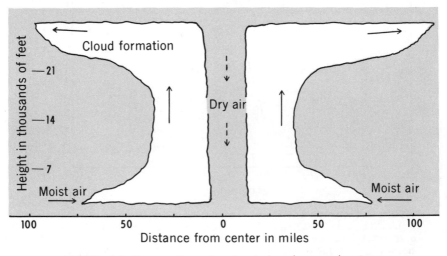

FIGURE 6.3 Cross-section of a tropical cyclone or hurricane.

canes the maximum winds will be in the vicinity of 125 miles per hour but in extreme cases may reach 200-250 miles an hour. Hurricane velocity winds may extend over an area 200-300 miles in diameter and gale force winds (28 miles per hour) over 400 to 500 miles in diameter. Hurricane winds blow with such a terrific force that they are difficult to measure. No one knows how strong hurricane winds can get. Often the winds are so strong that they blow anemometers and other instruments away. Special anemometers, reinforced against hurricanes, have measured such winds at velocities greater than 160 miles per hour. While the winds within the storm are of very high velocity, the storm moves at a relatively slow speed. They average only 10-20 miles per hour.

Very heavy precipitation is associated with hurricanes. Fifty cm (20 inches) over a forty-eight hour period are relatively common. One typhoon produced a total of over 100 inches during a forty-eight hour period in the Philippines. The precipitation seems to occur in bands away from the center, and some of the heaviest rains on record in low latitudes have been associated with these storms. Radar observations show that 300-400 miles ahead of the center of the storm there are often fairly well-defined lines of thunderstorms which move with the storm.

Once the hurricane begins to form, moist air from all sides converges toward the storm center, and condensation supplies heat, which is the energy needed to develop the storm. A constant supply of water vapor is essential for the formation and existence of the storm.

The storm moves slowly at first, usually moving from east to west in low latitudes. As it gains strength the speed increases and its path gradually curves poleward. As long as the storm remains over warm water it can grow in intensity. Part of the reason a hurricane will travel so far north along the east coast of the United States is because it follows the warm Gulf Stream. When the storm moves over the cold Labrador Current it dissipates rapidly. The storms will dissolve quickly over land due to increased friction and cooling caused by precipitation. The rainfall also cools the land surface which helps bring about stability, and as the storm moves over land the energy supply in the form of water vapor is cut off.

Tropical cyclones are found only in certain regions. They originate over the tropical oceans between latitudes 5° and 20°, and the greatest number form over the western sides of the oceans. Their absence along the Equator is attributed to the weakness of the Coriolis effect. There are six general regions of occurrence. They are the Caribbean Sea and Gulf of Mexico, the northwest Pacific from the Philippines to the China Sea, the Pacific Ocean west of Mexico, the South Indian Ocean east of Madagascar, the North Indian Ocean in the Bay of Bengal, the

North Indian Ocean and the Arabian Sea. Some hurricanes occur outside these regions but these are the areas of greatest frequency. It is of some interest that none have been noted south of the Equator in the Atlantic. Apparently the equatorial convergence zone does not migrate far enough south to provide the necessary convergence. Hurricanes are distributed in a fairly well-defined pattern, not only geographically, but in reference to time. The peak frequency corresponds to the period of highest sea temperature and the time of the maximum poleward displacement of the convergence zone. Thus late summer and early fall are the seasons of maximum occurrence.

The Tropical Rainy Climates (Af)

The tropical rainy climate is defined in terms of the high frequency of precipitation and fairly high temperatures throughout the year. The weather is a product of the maritime tropical air which prevails in these regions plus the weather associated with the thunderstorms and easterly waves.

This type of climate is typically found astraddle the Equator and extending poleward from 5° to 10°. The latitudinal width tends to be wider on the east side of the continents than it is on the west side due to the prevailing easterly winds and warmer water. This type of climate is restricted to low elevations, usually below 3500 feet; at altitudes greater than this adiabatic cooling lowers temperatures substantially. There are three major areas where this type of climate is found. They are in the Amazon Valley of South America, the Central Congo Valley, and parts of the East Indies. The area of the tropics which has these characteristics is much smaller than originally thought. Most of the tropics experience a pronounced seasonal regime of precipitation. Part of the reason for the restricted area of this climate is that the region must lie in a position within the tropical convergence zone all year, and this is not likely to happen over an extended amount of land because the convergence zone migrates with the sun.

Insolation is more uniform here than in any other climatic region. Intensity of radiation is high all year since the vertical rays of the sun move some 47° but are never more than 23°30′ from the zenith. Radiation reaching the surface would be much greater were it not for the high moisture content of the air. There is little variation in the length of the day through the year due to the intersection and bisection of the circle of illumination and the Equator.

Monthly mean temperatures as well as daily mean temperatures average from 26° C to 30° C (78° F to 86° F). In the United States, Key West has the highest average of any city with a mean annual temperature of 25° C (77° F). The extremes occur when humidity drops and there is the least cloud cover, permitting a maximum of insolation and earth radiation. The annual range in temperature is generally less than 2.8° C (5° F). Belem, Brazil, experiences a range from the warmest to coolest month of only 1.7° C (3° F).

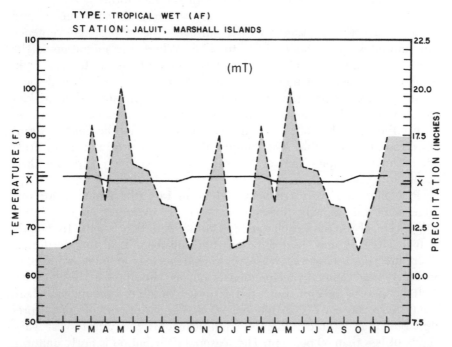

FIGURE 6.4 Climograph of a station in the humid tropics. The mean monthly temperature and precipitation (X̄) is plotted near the center of the vertical axis to make it easier to compare this climograph with the others to follow and also to better show the seasonal fluctuation. The temperature is shown in the solid line and precipitation in the dashed line. A two-year sequence is shown for the purpose of showing the periodicity of the climatic elements.

The main factors controlling the temperature are (1) the equal or nearly equal periods of daylight and darkness; (2) the nearly vertical radiation angle; and (3) the constant high humidity. The first two factors keep the seasonal temperature uniform; the latter one operates to keep the diurnal temperature even. The diurnal range in temperature exceeds the annual in the tropics and is usually less than 8.3° C (15° F) in the humid tropics.

The equatorial trough of low pressure is consistent but not particularly deep. The pressure usually remains slightly below the average for sea level but the trough is sufficient to bring about convergence most of the time. Occasionally weak easterly waves appear which deepen the trough somewhat. Winds are predominately from the eastern half of the compass rose. They vary from northeast to southeast depending on the actual location and season. Velocities at the surface tend to be low, as the pressure gradient is low and the vegetation at the surface retards the airflow. Occasional high winds occur with thunderstorms.

Moisture conditions are controlled by the tropical convergence and by the convective heating over the surface. The tropical convergence zone migrates from about 15° N to 5° S. Where the migration is the greatest the rainy region is the smallest. Humidity tends to stay high, being over 50 per cent most of the time. As humidity is high and there is convergence occurring, partly cloudy skies predominate and clear skies are infrequent. The cumulus forms are the most common, building up during the daylight hours into alto-cumulus and cumulo-nimbus forms when the air is unstable. The cumulo-nimbus clouds do not go to great heights in the tropics, and as a result hail is extremely rare but does occur on occasion.

Precipitation is very frequent and is, in fact, the major factor distinguishing this climate. Precipitation can be expected on more than 50 per cent of the days each month. The precipitation is most frequently of high intensity and short duration resulting from local thunderstorms. There are times, however, when prolonged periods of intensive rainfall occur. Average annual precipitation is not as important a consideration as frequency of precipitation, and is quite variable, ranging anywhere from 150 to 250 cm (60 to 100 inches) a year. An aspect of annual precipitation that is of considerable significance is the year-to-year variability of less than 50 per cent. The seasonal distribution is fairly uniform. While some months have more rainfall than others, all months have an average of 6 cm (2.4 inches) or more. Most places will average over 7.5 cm (3 inches) per month and many 10 cm (4 inches). Early morning fogs and heavy dews are both common in the rainy tropics. Since humidity is so high in the daytime, when cooling takes place at night the dew point is reached while temperatures are in the range from 18° to 25° C (65° to 75° F).

The Tropical Seasonal-Convergence Climates (Am, Aw, As, BSh)

The distinguishing characteristics of these areas are the year-round high temperatures and pronounced wet and dry seasons. They are found marginal to the tropical convergence climates and in most cases poleward from them. They occupy much of the area from the boundary with the tropical convergence region to the Tropics of Cancer and Capricorn.

TABLE 6.1
Climatic Data for Selected Tropical Stations

	J	F	M	A	M	J	J	A	S	O	N	D	Yr.
Jaluit, Marshall Islands													
T	78	79	79	79	80	80	80	79	79	79	78	78	79
P	10.8	7.9	9.8	10.8	10.7	8.7	6.3	8.9	8.4	14.8	15.7	13.2	125.9
Saigon, South Vietnam													
T	79	81	84	86	84	82	82	82	82	81	80	79	82
P	0.9	0.1	0.3	1.7	8.3	12.6	11.1	11.0	13.3	11.1	3.7	3.1	77.2
Key West, Florida													
T	70	71	74	77	80	83	84	84	83	80	75	71	77
P	1.5	2.0	1.7	2.5	2.8	4.0	4.2	4.3	6.5	5.9	2.8	1.7	39.8
Ascension Island													
T	79	80	81	81	80	79	78	77	76	77	77	78	79
P	0.2	0.2	0.7	0.5	0.3	0.4	0.5	0.4	0.4	0.3	0.2	0.2	4.2
Aswan, Egypt													
T	59	63	70	78	85	90	91	90	88	82	72	62	77
P						Trace							
Manáos, Brazil													
T	82	82	81	81	81	81	81	83	84	84	84	82	82
P	9.8	9.1	10.3	8.7	6.7	3.3	2.3	1.5	1.8	4.2	5.6	8.0	71.3
Kisangani (Stanleyville), The Congo													
T	78	78	78	79	78	77	75	75	76	77	76	77	77
P	2.1	3.3	7.0	6.2	5.4	4.5	5.2	6.5	7.2	8.6	7.8	3.3	67.1
Karachi, Pakistan													
T	65	68	75	81	85	87	84	82	82	80	74	67	78
P	0.5	0.5	0.4	0.2	0.1	0.9	2.9	1.5	0.5	T	0.1	0.1	7.6
Freetown, Sierra Leone													
T	81	82	82	82	81	80	78	77	78	80	80	81	80
P	0.6	0.5	1.1	5.4	14.8	21.3	36.8	39.6	32.5	15.2	5.3	1.3	174.4

T = temperature in °F and P = the precipitation in inches.
°C = 5/9 (°F-32)   cm. = 0.4 (inches)

Their location in the general circulation is that zone which experiences the seasonal migration of the equatorial convergence and the subtropical divergence.

The temperatures are controlled by latitude and the shift of the semi-permanent pressure belts. The variation in the length of day through the year is not too significant, as the photo-period varies from eleven to thirteen hours. There is substantial variation in the angle of incidence as the sun ranges on the poleward margins up to 47°. The shift of the pressure belts, and thus a change from the maritime to continental air, has a pronounced bearing on the diurnal and annual extremes.

Temperatures in these areas average about the same as in the tropical convergence areas, but there is somewhat greater variation. Some areas experience higher average annual temperatures than are found in the tropical convergence zone. This is due to the greater variation in radiation intensity associated with these areas.

Annual range in temperatures is somewhat greater, being between 2.8° C and 8.3° C (5° F and 15° F). This is a result of the dry season radiation conditions. The diurnal range of temperature during the wet season is quite comparable to that of the tropical convergence climates.

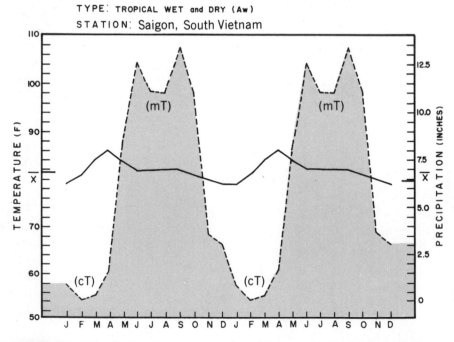

TYPE: TROPICAL WET and DRY (Aw)
STATION: Saigon, South Vietnam

FIGURE 6.5 Climograph of a representative station in the tropical wet and dry climate.

The dry season diurnal range is somewhat greater, averaging 11° C to 17° C (20° F to 30° F). As in the other tropical climates, the diurnal range exceeds the annual range. The frost-free season is continuous here, although temperatures occasionally will drop into the forties (°F) during the dry season on the poleward margin of the zone.

The really distinguishing feature of the tropical wet and dry areas is the pronounced seasonal moisture pattern. Atmospheric humidity during the high sun season is quite high and during the dry season it is very low. This is reflected in cloud cover. During the season of convergence clouds are present much of the time and clear skies are infrequent. The low and middle forms of cumulus clouds are most prevalent. Dew-point conditions change according to the seasons. During the high sun period the dew point will range in the 60s or even in the low 70s (16° to 22° C). But during the dry season dew points in the 30s or 40s (0° to 7° C) are more common.

Average annual precipitation is extremely varied and cannot be used as a criterion for distinguishing the region. On the dry margins it drops to less than 10 inches and on the wet margins, where topography is favorable for orographic intensification, the totals run over 400 inches. It is in fact quite likely that the highest annual average precipitation totals to be found on the earth occur in this tropical zone of seasonal convergence. Annual variability is higher here than in the tropical convergence region because the precipitation total in any one year is subject to the extent of migration of the equatorial convergence zone and the duration of the migration. The further an individual point is from the heart of the convergence belt the less the annual total is, the shorter the rainy season is, and the greater the annual variability is.

The seasonality increases poleward as might be expected. In a traverse away from the Equator, the low sun precipitation begins to decrease first. The high sun precipitation remains as high as in the tropical convergence zone. Midway between the equatorial convergence zone and the tropical desert the low sun precipitation drops to near zero. Continuing poleward, the high sun precipitation declines until it is no longer sufficient to support woody vegetation. The tropical steppe begins here. Grasses dwindle as high sun precipitation decreases. At the point where precipitation becomes insignificant (grass cannot grow) the tropical desert begins. The rainy season is concurrent with the high sun and presence of the intertropical convergence zone. The dry season is a product of rather stable desert air originating from the divergence in the subtropical high pressure zone.

This climate has the most pronounced seasonality of precipitation of any of the climatic types. An example of the extremes that are reached is found in Asia: Rangoon, Burma has a three-month low sun average of one inch (2.5 cm) and a three-month high sun average of 75 inches (190 cm). This increases northward to Akyab where the same amounts are 1.5 inches and 170 inches (3.8 cm and 432 cm). The maximum in the Asian region is reached at Cherrapunji where in two winter months they receive an average of only 1 inch of precipitation but in two summer months 211 inches (536 cm). Cherrapunji has recorded a five-day total of 150 inches (381 cm), a one month total of 366 inches (930 cm) and a yearly total of 1042 inches (2647 cm).

The Köppen classification subdivides the tropical seasonal rainfall climates into four subgroups. These are the monsoon (Am), winter dry (Aw), summer dry (As), and the tropical steppe (BSh). These categories all exhibit a rainy season and a dry season, but the dry seasons have difficult lengths or occur at different times of the year. The monsoon (Am) has a short dry season, but precipitation is abundant during the rainy season. There is enough rainfall during the rainy season to sustain forest growth during the dry season, and for this reason the climate is often considered as a part of the tropical rainy climate. The summer dry (As) climate is distinguished from the other seasonal rainfall climates by the occurence of the precipitation during the low sun period of the year. The area which has this kind of distribution of rainfall is not very extensive. The winter dry (Aw) and steppe (BSh) differ climatically in terms of the length of the dry season. The steppe has too short a rainy season to support tree growth to any extent.

In contrast to the fairly widespread area of the tropics which experiences a winter dry season is the summer dry (As) climate. It is limited in extent and the best known area is that along the east coast of South America from approximately 10° to 20° south. Here in eastern Brazil is an area which receives most of the annual precipitation in the winter half-year (April-October). It is parallel to the coast and extends only a short distance inland. The rainfall totals are generally over 30 inches (77 cm) a year and about 3/4 of this occurs in the winter months. In the winter months, Cape São Roque and the highlands inland are generally responsible for keeping the equatorial convergence zone from forming over the summer dry area, and hence less precipitation occurs than in the surrounding areas. During the winter months, however, a variety of weak storms produce frequent precipitation.

Altitude has a marked effect on temperatures in the tropics as elsewhere. Where elevations reach several thousands of feet temperatures may be lowered enough so that the site is no longer tropical in temperature (annual average less than 18° C). Usually the same seasonal precipitation and temperature pattern will exist at these slightly higher elevations as at sea level but the temperatures will average below those of the lower sites. This kind of upland climate (Cw) of the tropics exists in areas of South America and Africa. It differs from the C climates of mid-latitudes in terms of the lack of a well-defined season of winter and summer temperatures produced by polar and tropical air masses.

The Tropical Dry Climates (BWh)

The heart of the tropical dry climates (the tropical desert) lies near the Tropics of Cancer and Capricorn on the west sides of the continents. Most of the tropical deserts are crossed by both the tropic and the thirtieth parallel. This climatic region occupies more land space than any other type of climate. Specific areas in which this type of climate is found are northern Africa, Arabia, Pakistan, Iran, India, southwestern United States, northern Mexico, Australia, and parts of South America.

Solar intensity tends to be very high here due to the lack of both cloud cover and atmospheric moisture. There are times when blowing dust obscures and absorbs much of the radiation near the surface, but overall radiation is high even in the low sun season. Clear days are the rule and cloud cover is low. Clouds are not completely absent as small fair weather cumulus clouds occur. However, the stability produced by subsidence and the low humidity reduces cloud cover to a minimum.

The highest average annual temperatures of any climate are found in the tropical deserts. They will vary from 29° to 35° C (85° to 95° F). At Lugh Ferrandi, in Somalia, the temperature averages 31° C (88° F) for the year. This is considerably higher than in the other tropical climates. The major factor controlling the annual average temperature is latitude, and the stations with the highest averages are those closest to the equator. High sun daily and monthly averages are in the nineties (°F) and extremes will reach above 49° C (120° F). The highest temperature yet recorded is 58° C (136.4° F) at El Azizia, near Tripoli in northern Africa on September 13, 1922. In the United States the highest recorded temperature is 56.7° C (134° F) in Death Valley, California.

Low sun averages are below those of other parts of the tropics. Radiation at night cools these areas so rapidly that averages are in the sixties and seventies (°F). The lower temperatures of the low sun season give the tropical deserts the highest annual range of the tropical climates. Aswan, Egypt, located on the Tropic of Cancer, has an annual range of 19° C (34° F). Coastal stations where there is some marine influence will have lower annual ranges than the interior stations. The annual range for the station is controlled by latitude and atmospheric humidity.

Diurnal ranges in the deserts are the highest of any of the climates and they greatly exceed the annual range in the other tropical climates. The average diurnal range is from 14° to 25° C (25° to 45° F) but on occasion they are much greater than this. At Birmilrha in the Sahara south of Tripoli the temperature went from 37° C (99° F) down to —.5° C (31° F) in one day for a range of 37.5° C (68° F), and at the oasis of In-Salah in the Sahara, on one occasion, the temperature dropped from an afternoon high of 52° C (126° F) to –3° C (26° F) the following morning for a range of 55° C (100° F).

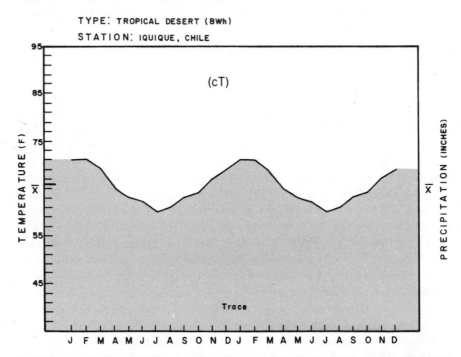

FIGURE 6.6 Climograph of a representative station of a tropical desert. The temperatures are somewhat lower for this station than for many tropical desert stations due to its location near the ocean.

The frost free season is twelve months except along the poleward margins where black frosts occasionally occur. Black frost occurs when the dew point is below freezing, but too low for frost to develop and the air temperature drops slightly below freezing. Plant tissues may be damaged even though there is no visible frost.

Humidity is very low throughout the year. Relative humidity averages 10-30 per cent in the interior areas and slightly higher in coastal locations.

The relative humidity has been measured as low as 2 per cent. With relative humidity as low as it is, cloud cover is also slight. The Sahara averages 10 per cent in winter and 4 per cent in the summer.

Precipitation is quite low in amount and very sporadic in distribution, both temporally and spatially. Low precipitation is the basic characteristic of these climates and in fact becomes a meaningless statistic in many cases. Arica, in Chile, averages 5 mm (.02 inch) of precipitation per year. Iquique, also in Chile, experienced a fourteen-year period with no rainfall and Wade Halfa in the Sahara Desert experienced a nineteen-year period with no rainfall. This characteristic of high variability is widespread. One rain may bring 12 to 25 cm (5 to 10 inches) of precipitation and then a period of several years may occur in which no precipitation at all is recorded. Variability of rainfall from year to year is likely to exceed 50 per cent. One station in the Thar Desert with an annual average of 12 cm (5 inches) received 86 cm (34 inches) in two days. At Dakhla, Sahara where the average is 10 cm (4 inches), one period of eleven years elapsed with no rain.

Throughout the tropical deserts elevation has an influence on precipitation. Areas of hills and mountains will receive more rain than lowland areas. Some high mountain ranges in the deserts have clouds overhead a great share of the time and enough precipitation to support green vegetation all year.

A characteristic of moisture in the deserts (and a characteristic which has been used to define deserts) is that potential evapo-transpiration exceeds precipitation. There is a fairly close relationship between temperatures and evapo-transpiration rates. In some places potential evapo-transpiration may be 25 to 50 times precipitation. In the tropical deserts seasonal distribution becomes meaningless. As can be readily determined, however, there is a greater frequency of precipitation on the equatorward margins in the summer time and on the poleward sides of the deserts in the winter. The precipitation that does fall is primarily convective on the equatorward side, with only occasional cyclonic precipitation. As latitude increases the frequency of cyclonic precipitation increases also.

The basic controls in the tropical deserts are the upper air stability, and subsidence. The deserts are enhanced or strengthened by being in interior locations away from a moisture source, being in the lee of a mountain range, or being on a coastal area where cool offshore currents exist. Seasonal weather is associated primarily with temperatures. With the low sun season come slightly cooler days and much cooler nights.

Severe storms are not frequent in the deserts. The main reason being the lack of moisture to supply the energy. The most often-seen desert storm is the *dust devil*. Dust devils are vortices in clear air with low humidity. They are the result of high surface temperatures and an air-flow with some curvature in it. A common feature of the air over deserts is a steep lapse rate and instability in the air near the surface. The dust devil results when there is an intense thermal, and air moves in at or near the surface to replace the rising air. As the air moves in toward the thermal the radius of curvature decreases and velocity increases. Dust devils are usually observed by the debris they carry and have been known to reach a height of nearly a mile. They will sometimes sustain velocities high enough to blow down shacks or blow screen doors off their hinges. These dust devils are very common in desert areas and often a goodly number may be visible at one time.

Along coastal sections of the tropical deserts a rather distinctive atmospheric condition sometimes prevails. These desert areas are characterized by cool temperatures, shallow temperature inversions, cold water offshore, considerable fog and stratus cloud cover, but no rain. Examples of such areas are coastal portions of lower California, Ecuador and Peru, and West Africa (Sahara and Namib). Examples of the humidity and precipitation conditions are given for Iquique and Caloma in Peru. At Iquique the relative humidity averages 81 per cent in August and the rainfall averages only 2.8 cm (1.1 inches) per year. During a twenty-year period, fourteen had no measurable precipitation. Caloma averages 48 per cent relative humidity in August and no measurable rainfall has ever been recorded. These areas have extremely low precipitation, even for desert areas, and much higher percentage of cloud cover. Humidity often hangs in the 80-90 per cent range and fog results. The high humidity is a product of the warm air moving over a cold offshore current. This condition does not carry very far inland as the air is rapidly modified by the ground surface.

Chapter 7

The Mid-latitude Climates

In mid-latitudes the primary circulation is quite different from that of the tropics. In mid-latitudes there is more horizontal air movement near the surface than is generally the case in the tropics. The subtropical high and the polar high act as major centers of subsidence and source regions for air moving toward the convergence zone centered in the latitudinal range of 50° to 60°. These two source areas are very different in character. One is a very warm area and the other quite cold. While this is sufficient to produce different characteristics of airflow into the mid-latitudes there are often substantial moisture differences also. Thus while in the tropical convergence zones converging air is quite similar in temperature and moisture, this is not the case in mid-latitudes. The result is a relatively large amount of turbulence, and the zone of convergence in the subpolar latitudes is the stormiest region of the earth's surface. The net circulation in mid-latitudes is from west to east, and is fairly strong.

The secondary circulation consists of a number of storm types, one of which is essentially a mid-latitude phenomenon, while the others are occasionally found in the tropics as well. The elements of the secondary circulation of the mid-latitude areas are the mid-latitude cyclone, the anticyclone, the thunderstorm, the tornado, and the occasional tropical cyclone (hurricane).

The boundary between two unlike air masses is often referred to as a *frontal surface,* and the line of intersection between the frontal surface and the ground surface is designated as a *front.* The leading or advancing edge of a cold air mass is called a cold front. As a mass of cold air enters the middle latitudes from polar regions, its front penetrates under the warmer air present. The cold air remains closer to the ground because it is denser and heavier than the warm air it is displacing. As the cold air continues to move under the warm, the force of momentum of the cold air behind it forces the warm air aloft. When a fast moving cold front displaces moist unstable air, the weather at the surface may become quite turbulent with gusty winds and intense precipitation lasting for a short period of time. The weather clears rapidly as the front passes by, the winds shift in direction toward the pole and temperatures drop. The passage of the cold front is often accompanied by rising atmospheric pressure and decreasing relative humidity. A slow moving front displacing stable warm air will produce slow clearing of the weather without any precipitation.

A warm front develops when the leading edge of a mass of warm air displaces colder air at the surface. Its lighter density prevents it from penetrating a cold air mass from underneath. Instead, it overruns the cold air. The weather associated with the warm front is more widespread and persistent than cold front weather. The air is lifted more slowly and as it cools clouds form which may exist as much as 1000 miles in advance of the warm front. The forward movement of a warm front can be observed long in advance of the surface front by gradually thickening and lowering clouds and development of rain or fog near the front. As the warm front passes, the temperature rises, wind shifts, precipitation stops, and cloudiness decreases.

Two other types of fronts are associated with cyclones, the *occluded front* and the *stationary front*. When a cold front overtakes a warm front, the air in the warm sector of the low pressure system is lifted. Cold air from the rear of the cold front will then meet with cold air in front of the warm front. It is possible that there will be a difference in temperature between the two cold air masses. If the air mass in the rear of the cold front is colder, it will displace the slightly warmer air by lifting it. This is a *cold front occlusion*. If the air in front of the warm sector is colder, it will lift the air in the rear of the cold front upwards. The front then is a *warm front occlusion*. If the frontal separation between two air masses remains in approximately the same position on the surface, it is a *stationary front*. Occluded fronts usually bring poor weather conditions, as the air is lifted continuously over an area and prolonged precipitation may result. Precipitation is likely to take place in these cyclonic systems due to the extensive lifting of air and subsequent cooling. Thunderstorms may produce heavy rains of short duration, but cyclonic rainfall tends to be of lower intensity and of longer duration. The precipitation may last for a period of days and cover several hundred square miles.

Although cyclones do not follow definite tracks around the world they do travel over certain paths more frequently than others. It can be de-

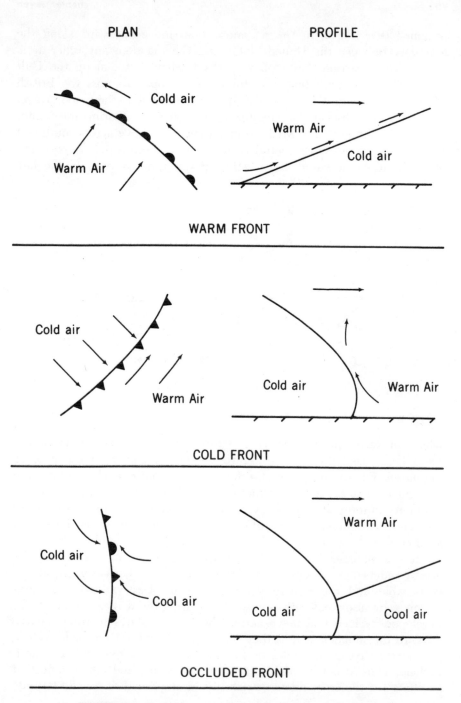

FIGURE 7.1 Most common types of atmospheric fronts.

termined that in North America more lows move eastward along the boundary between the United States and Canada than any other area. In the winter season, lows follow paths further south, along the Gulf Coast. The main path that lows follow in Europe is across the British Isles and northern Germany. In the winter months they, too, travel further south, crossing the Mediterranean countries. During the winter some of the more intense Mediterranean cyclones move across southwest Asia. Few cyclones cross central Asia in winter because of the strength of the anticyclonic circulation existing at this time of year. In the late

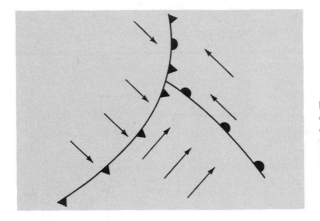

FIGURE 7.2 A plan view of a mid-latitude cyclone showing the distribution of fronts and associated direction of airflow.

winter or early spring these large storms may cross the center of the Asian land mass, producing blizzard conditions known in the area as the *buran*. For the most part the lows in the Southern Hemisphere are stronger than those of the Northern Hemisphere. The air mass contrasts are more uniform through the year in the Southern Hemisphere due to the almost continuous water surface in the latitudes in which the storms are found.

In the summer the intensity of the cyclones and anticyclones is greatly reduced. As a consequence, the summer weather is somewhat less changeable and atmosphere disturbances are less violent. The anticyclones are also weaker than in winter. During the winter season a well-developed cyclone will move across the United States in three to five days. In the summer months they move more slowly. After each cyclone passes there comes a period of fair weather associated with the anticyclone. Thus, it is the combination of the cyclone and anticyclone that accounts for the tremendous variability in the weather in mid-latitude areas.

The tornado is the most intense vortex that occurs in the atmosphere. It is a converging spiral of air with wind speeds estimated at several hundred miles an hour. It is the most violent of atmospheric storms but it seldom gets much larger than a mile in diameter. The direction of rotation is generally dictated by the Coriolis force. The tornado, like the thunderstorm and hurricane, depends upon moisture as an energy supply, and also like the thunderstorm and hurricane, most tornadoes occur in maritime tropical air. The storm is readily visible in the form of funnel or post-shaped clouds of condensed moisture hanging from a cloud. The cloud may vary widely in thickness, and sometimes it may be larger at the bottom than at the top. Most often it is gray in color due to condensed water vapor. As the tip contacts the ground the appearance may be changed by dirt and debris picked up. In winter they have been observed to touch down on a field of snow and become a brilliant white.

The storm develops extremely low atmospheric pressure in the center. The drop in pressure is so rapid and of such a great amount that often barographs have not been able to withstand the sudden pressure decrease. Also they are often destroyed by the winds associated with the storm. The greatest known drop in pressure occurred in Minnesota in 1904, when the pressure dropped to almost 813 mb (24 inches). Wind velocities in these storms have only been estimated. Some estimates have gone as high as 500 miles per hour.

Tornadoes occur most frequently in central and southeastern United States. Every state in the United States has experienced tornadoes but their occurrence in regions north of 45° and west of the Rockies is infrequent. Nine states in the United States report an average of over five tornadoes per year. They are Kansas, Iowa, Texas, Arkansas, Oklahoma, Missouri, Alabama, Mississippi, and Nebraska. They occur on the other continents but are relatively rare except in Australia, where they are almost as common as in the United States.

The temporal distribution of tornadoes in the United States is similar to that of thunderstorms. The diurnal pattern consists of a maximum concentration in a two-hour period between 4:00 P.M. and 6:00 P.M., with approximately 25 per cent of the tornadoes occurring at this time. About two-thirds of all tornadoes occur in the six-hour period from 2:00 P.M. to 8:00 P.M. The annual pattern contains a maximum in the months from May to September, but they occur in all months. The number of tornadoes which occur varies a great deal from year to year. In the United States in the period from 1916-1950, an average of 150 confirmed tornadoes a year occurred. During the same time the number varied from 62 to a high of 220 a year.

Tornadoes are almost always associated with thunderstorm activity. Although they may be produced from a variety of conditions, the best conditions for tornado formation are found ahead of a cold front. They form from the collision of a mass of warm, very moist air with cooler drier air from poleward regions. Extreme turbulence develops along the air-mass boundary and eddies occasionally develop into strong whirls through which warm air escapes upwards. If the air is extremely unstable the convergence intensifies and the storm forms. Perfect conditions exist in the states already named during the spring, when warm, moist Gulf air meets a cold front along a north-northeast, south-southwest line.

Mid-latitude Convergence Climates (Cf, Df)

The mid-latitude convergence climates are found associated with subpolar lows over continental areas and spreading equatorward on the eastern sides of the continents. They spread across the land masses in the vicinity of 60° N and S and reach equatorward as far as 25° or 30° along the eastern margins of the land masses. This type of climate in North America extends from the Pacific Coast of Canada eastward in a crescent to the Atlantic Coast. In the United States it includes that area lying east of the Mississippi River and the first tier of states to the west of the river. In Eurasia this climate extends from the offshore islands of Great Britain and Ireland eastward into the Soviet Union. It is also found along the Pacific Coast of the Orient in latitudes 30° to 60°. In the Southern Hemisphere similar climatic regions are found in South America, Africa, and Australia. In South America there are two areas, geographically separated. One is in the south of Chile, including the southern end of the central valley, and the other portion includes the Pampas of Argentina and Uruguay and extends southward to include part of Patagonia. The area in Africa is extremely limited, consisting of a small area on the very southeast tip of the continent. Both Australia and New Zealand have sectors with this type of climate.

Temperatures must be discussed in terms of seasonal conditions. During the high sun season temperatures will average in the seventies (21° to 26° C) at most stations. Along the equatorward boundary the warm month temperatures may average in the low eighties (27° to 29° C), comparable to the rainy tropics. The warm month means drop slowly in a poleward direction. Only when latitudes greater than 45° are reached will the summer averages drop below 70° F (21° C). July averages differ only slightly from the Gulf Coast to the Canadian boundary.

The summer extremes exceed those of the tropical convergence climates. Whereas few stations in the tropical convergence climates have

TYPE: MID-LATITUDE WET (Cfa)
STATION: CINCINNATI, OHIO

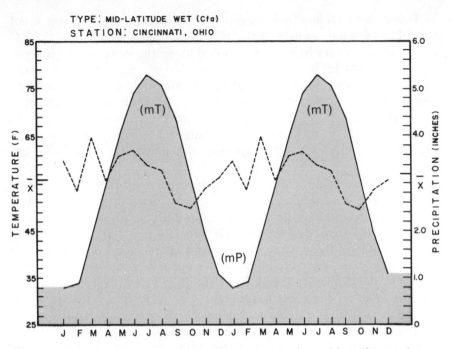

FIGURE 7.3 Climograph of a representative station in the mid-latitude wet climate.

recorded temperatures in excess of 100° F (37.8° C), most stations in the mid-latitude convergence climates have recorded maximums from 100° to 110° F (37.8° to 43.3° C). The summer diurnal range is typically from 8° to 11° C (15° to 20° F), again quite comparable with the tropical convergence climates. In the three summer months of June, July, and August, the temperatures of southeastern United States averages 2° to 3° F (1° to 2° C) higher than in Belem, Brazil, and the rainfall is about the same. The nocturnal temperatures are quite often within a few degrees of the diurnal temperatures, as the nights are short. The summer temperatures of the mid-latitude convergence climates are controlled mainly by the high solar intensity, long solar duration, and high atmospheric humidity.

It is the low sun season which sets the mid-latitude convergence zone apart from the tropical convergence climates. In the low sun season the solar intensity is low and the solar duration is also short. Temperatures reflect the rapid change from the high sun conditions. While temperatures averaging in the fifties (10° to 15° C) are the rule on the equatorward margin, the low sun averages drop rapidly to below 0° F (−17.8° C) in interior stations in higher latitudes. There is a rapid decrease in the mean winter temperatures as one progresses poleward.

The winter extremes have a steep gradient poleward also, but it is not as marked as is the gradient of the average temperatures. Near −17.8° C (0° F) temperatures have been recorded along the equatorward margin, and lows from −32° to −46° C (−25° to −50° F) characterize the more northerly locations. Winter diurnal ranges are slightly greater than the summer ranges as the relative and absolute humidities are somewhat lower. The annual range increases poleward from about 8° C (15° F) to 38° C (70° F) at the extreme. The winter-time controls are in essence, low solar intensity, low solar duration and moderate atmospheric humidity.

Precipitation in these regions is fairly evenly distributed throughout the year. The annual total is quite variable, depending upon latitude and continental position. It varies from as little as 51 cm (20 inches) upwards to 178 cm (70 inches). Of the cities in the United States, Mobile is among the wettest with a mean annual total of 173 cm (68 inches) per year. The variability of the annual precipitation is similar to that of the tropics; normally less than 20 per cent. The seasonal distribution is fairly even, but some areas show a tendency toward either a summer or winter maximum. Frequency of precipitation also varies. Bahía Félix, Chile, averages some three hundred and twenty-five days a year with rain, thus there is a 90 per cent probability of rain on any given day of the year. This represents an above-average frequency, however.

The precipitation in the mid-latitude convergence zones is much more varied in form than in the tropical convergence zone. During the summer and in equatorial margins, convectional rainfall is the primary form of precipitation. The southeast of the United States averages some forty to sixty days per year with thunderstorms.[1] The frequency of the thunderstorms decreases rapidly from south to north. Cyclonic precipitation prevails on the polar margins and is more frequent in all sections of the region in the winter. Hurricanes provide another mechanism for inducing precipitation and some of the heaviest rains along coastal areas are the result of hurricanes.

While the greatest share of the annual precipitation is in the form of rain, snow is a factor to a greater or lesser degree in all parts of the region. In the United States the mean snowfall varies from a trace along the Gulf to approximately 100 cm (40 inches) as an average through the Great Lakes district. Although the water equivalent of snow varies a great deal, there is an average equivalent of about 2.5 cm (1 inch)

---

[1]The greatest occurrence is in the vicinity of Tampa Bay (Florida), where approximately one hundred days a year have thunderstorms with severe lightning.

of water to 25 cm (10 inches) of snow. The heaviest snowfalls often occur late in the fall or early in the spring, when very strong contrasts exist in the maritime tropical air and maritime polar air.

Humidity is generally high. On occasion the humidity will drop substantially but usually for short periods of time. High humidity and high temperatures make the sensible temperatures of summer quite high. With high humidity there is a high frequency of cloud cover. Cumulus clouds predominate in the summer, with more stratus clouds in the winter. And as in the tropical regions summer radiation fogs are common.

The seasonal weather is distinctive between the two seasons. The summer is much like the rainy tropics. Temperatures are high, humidity is high, and convectional showers prevail. Summer storms that add variety are the tornadoes, thunderstorms, and hurricanes. Other changes that bring variety to weather are the heat waves and cold waves of summer. The heat waves are the result of eastward migration or flow of continental tropical air from the deserts. Though infrequent, they bring hot dry weather which will produce temperatures of over 100° F (38° C). The cold waves are the result of equatorward flow of summer continental polar air from the interior of high latitude land masses. Though referred to as cold waves they may not have temperatures below 10° C (50° F), but they represent much cooler air than is normal.

The winter season is subject to much greater variation than is the summer season. Cyclonic showers of rain or snow, blizzards, cold waves and January thaws all add variety to the weather.

The mid-latitude convergence climates are characterized by the distinct summer and winter temperature regime and frequent precipitation at all times of the year. Within the region which displays these characteristics there are differences in temperature which have lead to subdivisions of the climate. Köppen distinguishes between the areas which have mild winters (Cf) and those with cold winters (Df). A similar subdivision is made on the basis of summer average temperatures. The two categories are those areas with hot summers (Caf) and those with cool summers (Cbf). The latter type of region is found predominately on the western sides of the continents where summer temperatures are modified by the onshore winds from the oceans.

Mid-latitude Seasonal Convergence Climates (BSk, Cw, Cs)

The mid-latitude seasonal convergence climate is characterized by a marked seasonal pattern in both temperature and moisture. The general location for this type of climate (BSk, Cw) is the interior of the continents. Specific areas of the Great Plains of the United States and Canada,

TABLE 7.1

Climatic Data for Selected Mid-latitude Stations

	J	F	M	A	M	J	J	A	S	O	N	D	Yr.
Cincinnati, Ohio													
T	33	34	44	54	65	74	78	76	69	58	45	36	56
P	3.5	2.9	4.0	3.1	3.6	3.7	3.4	3.3	2.6	2.5	2.9	3.1	38.6
Dodge City, Kansas													
T	29	33	43	54	64	73	78	78	69	56	43	33	54
P	0.4	0.8	0.9	1.9	2.9	3.3	3.1	2.7	1.9	1.3	0.7	0.6	20.5
Los Angeles, California													
T	55	56	58	59	62	66	70	71	69	65	61	57	62
P	3.1	3.1	2.8	1.0	0.5	0.1	T	T	0.2	0.7	1.2	2.6	15.2
Winnemucca, Nevada													
T	29	34	40	47	54	63	71	69	59	48	38	30	48
P	1.0	0.9	1.0	0.9	0.9	0.7	0.2	0.2	0.4	0.6	0.7	1.1	8.5
Christchurch, New Zealand													
T	61	61	58	53	48	43	42	44	48	53	56	60	52
P	2.2	1.7	1.9	1.9	2.6	2.6	2.7	1.9	1.8	1.7	1.9	2.2	25.1
Verkhoyansk, U.S.S.R													
T	-59	-47	-24	7	35	55	60	50	36	5	-34	-53	3
P	0.2	0.1	T	0.1	0.2	0.5	1.2	0.9	0.2	0.2	0.2	0.2	3.9
Rome, Italy													
T	44	46	51	56	64	71	76	76	70	62	53	46	59
P	3.5	3.1	3.0	3.2	2.3	1.9	0.9	0.9	2.9	5.4	4.7	4.1	36.4
Valdivia, Chile													
T	62	62	59	54	50	47	46	47	49	53	56	59	53
P	2.6	2.9	5.2	9.2	14.2	17.7	15.5	12.9	8.2	5.0	4.9	4.1	102.4
Boston, Massachusetts													
T	27	28	35	45	57	67	71	69	63	52	41	32	49
P	3.4	3.1	3.7	3.7	2.9	3.3	3.3	3.3	3.5	2.8	2.9	3.4	38.9

T = temperature in °F. and P = precipitation in inches
°C = 5/9 (°F-32)   1 cm. = 0.4 (inches)

the steppes of Eurasia, and small areas of Australia, Africa, and South America.

Solar radiation follows the same pattern here as in the mid-latitude convergence climate. The one big difference in insolation is the result of lower atmospheric humidity. There are many more clear days and thus both insolation and earth radiation are at a higher rate.

High sun average temperatures differ little from those of the more humid areas at the same latitude as they range downward from 26.7° C (80° F) with increasing latitude. The high sun extremes are greater due to more frequent dry air masses. Whereas the extreme highs in the mid-latitude convergence climates are predominately from 38° to 43° C (100° to 110° F), in the seasonal convergence regions they will range up to 49° C (120° F) in the warmer sections.

The cool season temperatures differ more from those of the humid areas. The low sun season is also the dry season. This favors earth radiation, with the result that mean temperatures are usually a few degrees less than in the convergence climate and extremes from 6° to 11° C (10° to 20° F) cooler. It is this climatic region which experiences the

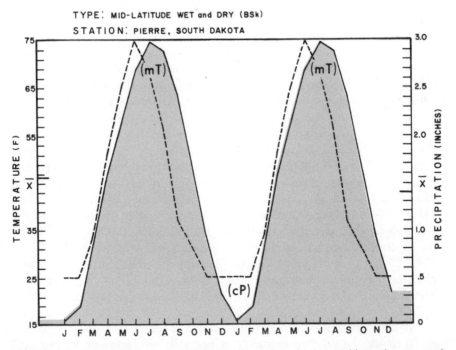

FIGURE 7.4 Climograph of a representative station in the mid-latitude wet and dry climate.

lowest temperatures of the Northern Hemisphere. In North America the temperature dropped to −62.8° C (−81° F) at Snag, Yukon, February 3, 1947, and at Tanana, Alaska −60° C (−76° F) was recorded in January, 1886. The lowest official temperature of the Northern Hemisphere was a −71.1° C (−96° F) at Verkhoyansk, U.S.S.R., and an unofficial −77.8° C (−108° F) was recorded at Oimekon, U.S.S.R.

As a result of less humid conditions and excessive earth radiation, the average annual range is up to 67° C (120° F) in the extreme cases and the absolute range up to 106° C (190° F). The diurnal range during the high sun season averages 11° to 17° C (20° to 30° F) and during the dry season ranges in excess of this are frequent. A winter characteristic of these areas is rapidly changing temperatures. At Browning, Montana, on January 23-24, 1916, the temperature dropped from 6.7° C (44° F) to −48.9° C (−56° F) for a total of a 55.6° C (100° F) variation in 24 hours. Such drastic changes are usually the result of the introduction of exotic air (usually polar but sometimes tropical), but adiabatically heated air will also cause equally sharp rises in the temperature.

These regions are further subject to the effects of a monsoon. During the high sun season the terrestrial heating causes the subpolar low to expand equatorward, forming a low trough or cell in the middle of the continent. The resultant convergence brings moisture on to the continent, producing a summer rainy season. During the winter, with excessive cooling, the subpolar low splits into cells over the oceans and a ridge of high pressure develops between the polar high and the subtropical high. This has divergence of either a subsidence or flow type, producing the winter dry season.

Winds are also a significant factor in the weather of the region. With the seasonal change in pressure is associated a seasonal change in wind direction. During the high sun season on-shore (on-land) winds prevail. During the winter there is a switch from 90° to 180° in prevailing wind direction. The degree of wind shift depends on the continental position. The further interior the station the greater the wind shift. Wind velocities are high as well as persistent. In the United States, Oklahoma City has the highest average wind velocity of any city, with a mean of 14 miles per hour. Chinook winds are a winter phenomenon that produce extremely rapid changes in temperatures. They occur in the Great Plains from Alberta to Colorado. Spearfish, South Dakota, observed a two-minute rise of 27.2° C (49° F) from −20° to 7.2° C (−4° to 45° F) January 22, 1943.

Atmospheric humidity and precipitation is seasonal in amount. During the summer the relative humidity is normally in excess of 50 per cent, while during the low sun season it will be below this amount. Cloud cover varies with humidity. Summer skies are partly cloudy with only a few days of clear skies. Winter is a season with little cloud cover and a high percentage of clear dry days. Precipitation decreases with latitude and distance from the sea. Although it is difficult to delimit annual precipitation most areas receive between 30 and 90 cm (12 and 35 inches). The annual amount is highly variable and apparently subject to long term cycles which are not as yet explained. Conjecture has placed the association with sunspot activity. Of the annual total, approximately 75 per cent falls during the summer half-year.

The storms most often associated with this climate are the tornado, thunderstorm, blizzard, and dust storm. Most tornadoes develop along a cold front and hence along a boundary between the mid-latitude rainy and seasonal rainfall climates.

One area with a seasonal rainfall regime (Cs) differs from the rest in that the rainy season comes in the winter rather than in the summer season. This type of regime is found on the western margins of the conti-nents between 30° and 40°, and normally doesn't spread into the conti-ents very far and is quite limited in extent. Specific areas with this type of climate are southern California west of the Sierra Nevada Mountains, including the Central Valley, central Chile, the Capetown district of South Africa, southeastern and extreme southwestern Australia, and the lands around the Mediterranean Sea. This climate is often referred to as the Mediterranean Climate.

The seasonal distribution makes this area rather distinct from the rest of the earth. Most locations experience a summer maximum of precipita-tion or at least have as much precipitation in summer as in winter. Here the winter maximum is very pronounced. For example, Santa Monica, California, has recorded only traces of precipitation in the three months of June, July, and August. The six month average for April to September is only 1 inch but the annual average is about 15 inches (38 cm).

Annual precipitation totals increase poleward and are generally low, being in the range from 25-65 cm (10-25 inches). Annual variability tends to be high, particularly on the equatorward margins. Humidity and cloud cover follow the same seasonal pattern. Because of the low humidity, and infrequent precipitation even in winter, these areas have a high percentage of sunshine throughout the year. It is the relatively mild and sunny winters which make these areas favorite winter resort areas.

The climate is a product of the subsidence associated with the sub-tropical high and the equatorward margin of the Westerlies. In the sum-mer months the strong divergence of the subtropics moves poleward over these areas bringing essentially desert weather. In the winter months

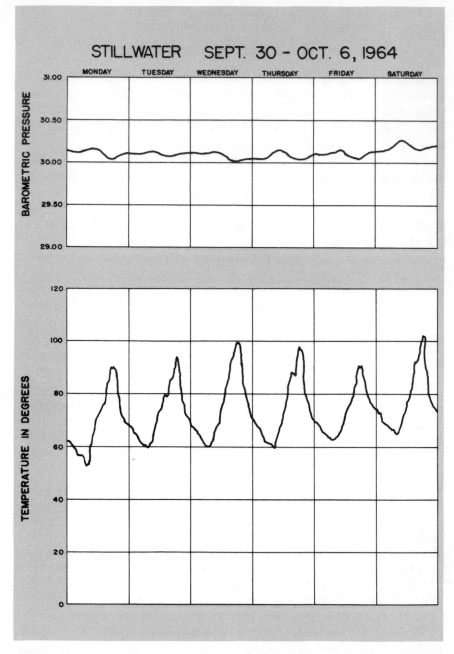

FIGURE 7.5 A week of temperature and pressure records at Stillwater, Oklahoma. It is typical of the kind of temperature conditions to be expected when storm systems are not present. The diurnal fluctuations represent the major changes in temperature when pressure and humidity are reasonably constant.

the anti-cyclonic circulation moves equatorward allowing the Westerlies to bring moisture into the area. The first fall rains are often torrential. Most of the precipitation is cyclonic and in the fall the land surface is extremely warm, which increases the instability of the air as it moves over the land. The occurrence of the precipitation in the wintertime and the nature of cyclonic precipitation makes the moisture far more efficient than in the other seasonal rainfall climates.

The temperature regime is closely associated with proximity to the sea. Coastal stations have cooler temperatures in the summer and warmer temperatures in the winter than interior stations. Capetown, South Africa has a warm month average of 21.1° C (70° F) and a cool month average of 12.8° C (55° F). Los Angeles averages 22.8° C (73° F) in August and 12.8° C in January. The extremes of temperature are also moderate for the latitude, and certainly moderate compared to the other mid-latitude climates.

Mid-latitude Divergence Climates (BWk)

The mid-latitude divergence climates (deserts) are located basically in the interiors of continents, although they merge with the tropical deserts on the west sides of the land masses. Specifically they are found in North America from southern Arizona northward into British Columbia between the Sierra Nevada Mountains and the Rockies. In Eurasia they are found imbedded in the trans-Eurasia cordillera, or lie on the flanks of these mountain masses. All of the central Asian countries contain substantial areas of desert within their boundaries. The Southern Hemisphere has a much smaller area of desert in mid-latitude locations. The fundamental reason for the diminutive size of these deserts south of the Equator is a lack of land area in the latitudes where the deserts would be most extensive.

Insolation in the mid-latitude deserts follows the same pattern as the other mid-latitude climates, since it is found in the same latitudinal range. High sun averages are the highest of any of the mid-latitude climates. The warm month averages will be from 6° to 8° C (10° to 15° F) higher than in the humid mid-latitude climates. The extremes of temperature are accordingly higher. During the summer, when the days are long and the sun high in the sky, temperatures will reach above 45° C (120° F). Death Valley in California has recorded an official shade temperature of 56.7° C (134° F), which is only 2° F below the highest sea level temperature ever recorded. The low sun averages compare with the humid climates of the same latitude. The mean annual range in monthly temperatures is from 22° to 28° C (40° to 50° F). The diurnal range averages 14° to 25° C (25° to 45° F).

Atmospheric pressure tends to remain fairly high during the winter but is very near normal sea level pressure or below during the summer. The pressure is not as important a factor in producing the mid-latitude deserts as it is in producing the tropical deserts. Flow divergence and sheer distance from the moisture source are the main factors.

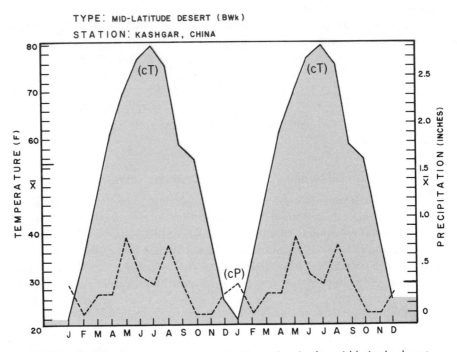

TYPE: MID-LATITUDE DESERT (BWk)

STATION: KASHGAR, CHINA

FIGURE 7.6 Climograph of a representative station in the mid-latitude desert.

This climatic area is the most difficult to place in the pattern of the general circulation of the atmosphere. As was indicated earlier, a number of factors are associated with the aridity of the mid-latitude deserts, each of which would be insufficient by itself to produce extreme draught, but when coupled with the other elements produces aridity. The mid-latitude deserts are not as arid as the tropical deserts, however.[2]

Humidity varies somewhat through the year with slightly higher humidity in winter than in summer. During the winter the subtropical high migrates equatorward, away from these areas, allowing occasional passage of weak lows associated with the subpolar convergence zone to

---

[2]This, of course, may not prove so when better data are furnished for central Asia. The Gobi may be as arid as any region on earth.

move through. These areas have the highest percentage of possible sunshine of any of the mid-latitude climates. Arizona averages 85 per cent possible sunshine, 94 per cent in June and 76 per cent in January. Precipitation totals are quite low. Phoenix averages 18 cm (7 inches) of precipitation per year, and Ellensburg, Washington, 23 cm (9 inches). Variability is high, as might be expected, averaging over 50 per cent. There is little seasonal pattern apparent, but what pattern exists is often a reflection of the seasonal pattern of the adjacent humid climates. Snow is part of the normal precipitation pattern. Phoenix receives traces of snow and Ellensburg averages 79 cm (31 inches) of snow in its total of 23 cm (9 inches) of precipitation. Moisture efficiency is very low throughout these regions. In some areas the potential evaporation exceeds the precipitation by as much as ten times or more. The highest potential evaporation in the United States is in the Desert Southwest.

Seasonal summer weather consists of heat waves, thunderstorms, hailstorms and tornadoes. In the winter cold waves and blizzards bring variety to the daily weather.

Chapter 8

Polar and Highland
Climates

The primary circulation in the polar climates is largely associated with the polar divergence, subpolar convergence, and the intervening air-flow. The three polar climates are located with respect to these three areas of the primary circulation. The polar deserts are associated with the divergence and subsidence at the poles, the polar marine environment is associated with the subpolar convergence over the oceans, and the polar seasonal precipitation climates are found on the marginal lands which experience several months of weather dominated by each system.

The polar climates are a product of control by polar maritime air masses and polar continental air masses. The polar climates are not without seasons, which are a result of the change between maritime and continental air masses and the variations in insolation which occur between winter and summer.

The Polar Convergence Regions (ET, maritime phase)[1]

The polar convergence climates are characterized by year-round low temperatures and high humidity. They are thus associated with the subpolar convergence, particularly where it exists year round. This places these regions in the high latitudes centered on the sixtieth parallel and mainly on island locations. Specific areas where this kind of region is found are the Aleutian Islands, the southern tip of Greenland to eastern Norway, extreme southern South America, and in a belt circling the Antarctic.

---

[1]Under the terms of the Köppen classification, a few of these regions might classify as Dfc or even possibly Cfc. The omission of a maritime phase of the ET climate is a serious deficiency of the Köppen system.

Solar intensity in the area is never high, due to the high latitude. In the summer months however, solar duration may be quite prolonged, with over twenty hours of daylight during several months. High humidity and cloudiness, however, tend to have a considerable degree of control over insolation and temperature. The percentage of possible sunshine is not high and clear days are rare.

The temperature reflects the low, even distribution of insolation and high humidity. The high sun averages vary from 0° to 10° C (32° to 50° F), and the high sun extremes are in the sixties (°F). The low sun averages are between −7° and 0° C (20° and 32° F), so the annual range in temperature is not very great. Low sun extremes are well above the −17.8° C (0° F) which is exceptional for a polar climate. The reason for the mild winter temperatures and low annual range is the marine location of these areas. The land areas which have this kind of climate are either islands or peninsular locations with open water nearby. As the ocean does not freeze in these areas it provides a constant source of heat for the atmosphere. The diurnal ranges are also quite low due to the high humidity.

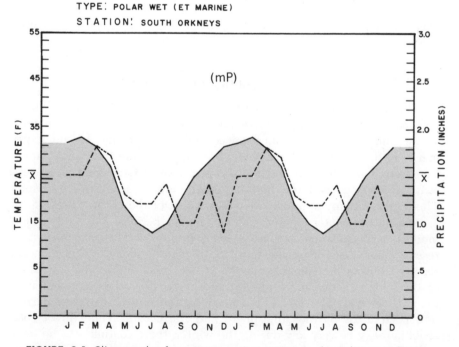

TYPE: POLAR WET (ET MARINE)
STATION: SOUTH ORKNEYS

FIGURE 8.1 Climograph of a representative station in the Polar wet climate.

A frost-free season is nonexistent. Frost can occur any day of the year, and these areas average over one hundred days per year with frost.

Pressure in these areas averages below normal. The result is year-round convergence. The subpolar lows tend to deepen in the winter with stronger pressure gradients. Moist conditions prevail all year and storms are frequent. In winter there is a much greater frequency of storms than in summer, with high winds and extreme variability of wind direction. This is the area which was named the Roaring Forties, the Furious Fifties, and the Shrieking Sixties, by the sailors of the sixteenth century.

Humidity is very high, averaging over 50 per cent all year. Cloud cover is also extensive, averaging 80 per cent during the winter and slightly less in the summer. Precipitation frequency is high in all areas with at least a 25 per cent probability of precipitation on any day of the year. The annual total is quite varied, however, as the intensity of precipitation varies with latitude and continentality. The totals vary from 38 cm (15 inches) up to 297 cm (117 inches). The annual variability of precipitation is among the lowest to be found anywhere. Snow is a common form of precipitation, but a ground cover of snow is short lived. Most snows are very wet and likely to last but a very short time.

This climate is distinctive in a number of ways. It is the cloudiest of all the climates with most of this area experiencing few clear days. No place can equal the South Atlantic zone for the frequency and severity of storms.

Polar Wet-and-Dry Climates (ET)

The basic defining characteristics of the polar wet-and-dry climates are year-round low temperatures and a seasonal precipitation regime. These climatic areas are found on coastal areas in high latitudes, primarily in the Northern Hemisphere. Specific areas are the North American Arctic coast, Iceland and Spitsbergen, coastal Greenland, the Arctic coast of Eurasia, and the high-latitude Southern Hemisphere islands. These areas occupy approximately 5 per cent of the earth's surface.

During the winter months, and in fact for periods as long as 8 or 9 months, earth radiation to space exceeds insolation. The polar areas represent major areas of escape to space of earth radiation. Since energy received during the summer months is not great due to the low angle of the solar radiation more radiant energy is released from polar areas than is received. This loss is balanced by the influx of energy carried into polar areas by the atmosphere and the ocean. While direct solar radiation may be very low during the winter, there continues to be some radiation received at the surface, part of which is in the visible range.

TABLE 8.1

Climatic Data for Selected Polar and Highland Stations

	J	F	M	A	M	J	J	A	S	O	N	D	Yr.
Lhasa, Tibet (12,900 feet)													
T	28	33	41	47	54	62	61	60	58	48	39	32	47
P	T	0.5	0-3	0.2	1.0	2.5	4.8	3.5	2.6	0.5	0.1	T	15.9
Quito, Ecuador													
T	55	55	55	55	55	55	55	55	55	55	55	55	55
P	3.2	3.9	4.8	7.0	4.6	1.5	1.1	2.2	2.6	3.9	4.0	3.6	42.3
Lauri Island, South Orkneys													
T	32	32	31	25	18	15	14	14	19	25	27	30	23
P	1.3	1.1	1.7	1.5	1.0	0.8	1.1	0.8	1.1	1.2	1.0	1.0	13.5
Angmagssalik, Greenland													
T	21	18	22	27	36	42	46	44	39	32	27	22	32
P	2.1	2.5	3.0	1.9	2.0	1.4	1.6	2.5	3.0	3.6	3.8	2.3	29.7
Spitsbergen, Island													
T	4	-2	-2	8	23	35	42	40	32	22	11	6	18
P	1.4	1.3	1.1	0.9	0.5	0.4	0.6	0.9	1.0	1.2	1.0	1.5	11.8
Little America, Antarctica													
T	19	7	-9	-21	-25	-28	-39	-38	-38	-15	2	19	-14
P					Insufficient data								
Eismitte, Greenland													
T	-43	-53	-40	-24	-4	4	12	1	-18	-32	-46	-37	-22
P					Insufficient data								
McMurdo Sound, Antarctica													
T	24	16	4	-9	-11	-12	-15	-15	-12	-2	14	25	1
P					Insufficient data								
Longs Peak, Colorado (8960 feet)													
T	23	22	26	33	41	51	55	55	48	39	31	24	37
P	0.7	1.2	2.0	2.7	2.4	1.6	3.6	2.2	1.7	1.7	0.9	0.9	21.6

T = temperature in °F and P = precipitation in inches
°C = 5/9 (°F, -32)   cm. = 0.4 (inches)

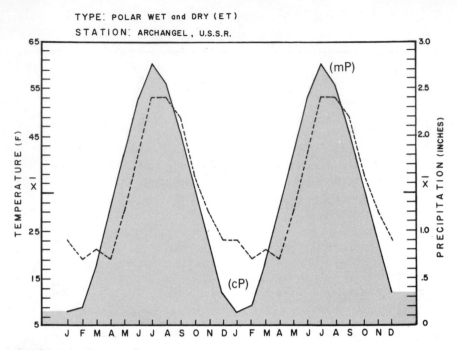

FIGURE 8.2 Climograph of a representative station in a polar wet and dry climate.

Even when the sun is below the horizon the reflection and refraction of the sun's rays produce twilight. Twilight extends the radiation limit up to 18° (astronomical twilight) beyond the actual circle of illumination. The auroras also provide light for the polar areas. As a result of these various sources of radiation, darkness is not as intense as is often believed.

Cloudiness varies through the year in the tundra. In the summer months cloud cover may average 80 per cent but during the winter it drops to around 40 per cent.

Temperatures vary considerably through the year. High sun averages are mainly in the range from 0° to 10° C (32° to 50° F) with the extremes going to 24° C (75° F). Although summer insolation is long in duration the greatest share of this insolation goes to evaporate water or melt snow and ice. The low sun temperatures are low and tend to stay low. Diurnal ranges in summer and winter are therefore extremely low. Low sun extremes are not as cold here as further equatorward. There is always some open water in the pack ice to modify the air temperature, as the Arctic Sea never freezes solid. The floes are from 10 to 100 feet thick and move from east to west with the Arctic current. The lowest

temperature yet recorded over the ice pack is −52° C (−62° F). Average annual range of temperature is from 22° to 33° C (40° to 60° F) and the absolute range from 42° to 56° C (75° to 100° F). There is no month in which frost does not occur.

Atmospheric pressure tends to change at a low rate from weak convergence to divergence with the seasons. Since pressure gradients are normally weak, this is not a stormy region. In fact, it is one of the least stormy areas of the earth's surface. High pressure dominates with cold, still, dry air, particularly in winter. Spring and fall are the seasons with most weather. Winds are light in velocity and variable in direction. The average velocity for the tundra is less than 10 miles per hour. In the Canadian Tundra, it is less than 6 miles per hour. High velocity winds are infrequent.

Atmospheric humidity is high during the summer months averaging from 60-80 per cent; during the winter it is less, averaging 40-60 per cent. Annual precipitation averages less than 25 cm (10 inches) for most stations and more than three-fourths of the annual total falls during the summer half of the year. Winter snowfall equals about 5 cm (2 inches) of water as the snow is dry and compact. A deep snow cover is absent over much of the tundra.

Polar Deserts (EF)

The polar dry lands are found in the high latitudes over land areas, largely on Greenland and the Antarctic continent. This climatic type is, surprisingly enough, one of the larger of the climatic regions. It occupies over 6,000,000 square miles.

Insolation is zero, or near zero, for half the year. During the summer months insolation is fairly high as the atmosphere is relatively transparent due to the low moisture and thin cloud layers when compared to mid-latitude locations. The incoming radiation is often quite diffuse as a result of the thin clouds, reflectivity of the snow, and cloud reflection. This diffusion produces the "no-shadows" effect of polar areas. The albedo of snow covered surfaces in the Antarctic averages 90 per cent. In mid-summer when the surface is roughened by melting, the reflectivity may drop to 65 or 70 per cent. As a result of this high reflectivity little of the radiation is absorbed. During a period of a few weeks in the summer there is heat added to the snow surface by radiation, but over the year there is a net loss of energy by radiation. There is little change in snow temperature with time and there is little upward transfer of heat through the snowpack. Since there is little energy added to the lower atmosphere from the snow surface the heat must be carried into the

region by the atmosphere itself. Convergence in the middle and upper troposphere provides the mechanism for advecting heat over the ice caps.

Empirical data is not abundant for the Antarctic or the Greenland ice cap. Continuous data collection in the Antarctic began essentially with the advent of the International Geophysical Year in 1957. During this 18 month "year" some 40,000 scientists and related workers from over 60 countries engaged in cooperative investigation of the earth and its environment.

Average annual temperatures are very low, all being below −17.8° C (0° F). Some examples are −30° C (−22° F) at Eismitte, Greenland; −49.4° C (−57° F) at Amundsen-Scott Station, Antarctica; and −57.2° C (−71° F) at Vostok, Antarctica. The seasonal temperatures reflect the variation in insolation, even if the sun never reaches more than 23°30' above the horizon. The low sun averages range from −31.7° to −73.3° C (−25° to −100° F). The lowest temperatures yet recorded at the surface of the earth were recorded in the interior of the Antarctic. At Vostok the temperature dropped to a record −87° C (−125° F) on August 25, 1958. In summers the temperatures range upwards toward the freezing mark but rarely reach it. High temperatures range from −20° C (−4° F) to above 25° C (77° F) on the coast. Beneath the ice, at depths

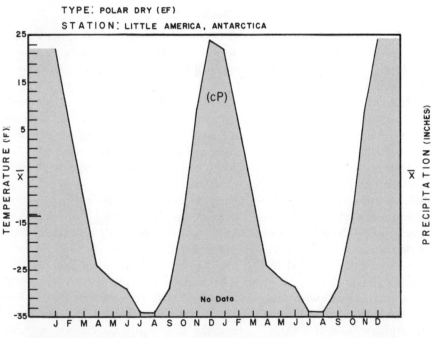

FIGURE 8.3 Climograph of a representative station in the polar desert.

of about 100 feet, snow temperatures have an average value of $-51°$ C ($-61°$ F). Over most of the Antarctic it does not get warm enough in the summer to melt the surface snow.

The cooling of the atmosphere by the snow surface produces a temperature inversion. The inversion persists and may grow to a depth of 1/2 mile or more and contain a vertical increase in temperature of as much as $30°$ C ($54°$ F).

Winds are a predominant factor in the weather of the polar deserts. Winds of a high enough velocity to move snow and produce blizzard conditions occur at Byrd Station about 65 per cent of the time and are of high enough velocity to produce zero visibility about 30 per cent of the time. A slight increase in wind velocity brings a substantial increase in blowing snow. The amount of snow moved by the wind varies as the fourth power of the velocity. It is unfortunate for those working in the Antarctic but the most accessible locations are also the windiest. Mawson's base at Commonwealth Bay experiences winds which are above gale force (28 mph) more than 340 days a year.

Strong gravity winds develop which blow off the land masses. These winds, *katabatic winds,* are the flow of cold dense air down a topographic slope due to the force of gravity. They result from the extreme cooling of the air over the ice. They may be strong enough to produce an airflow contrary to the normal pressure gradient. A topographic slope of about 0.2 per cent is about equal to the force of the average pressure gradient over the Antarctic. Where the slopes are steep the katabatic winds may regularly flow in a direction other than that which would be indicated by the pressure gradient. The strongest winds occur when the pressure gradient and topographic slope are coincident. The katabatic winds are the prevailing winds over much of the Antarctic, particularly where there is a steep drop from the interior highlands to the coast. Where these winds are fairly constant they will produce formations of ripples in the snow surface called *Sastrugi.* Strong winds and blizzards are often accompanied by rising temperatures, as the temperature inversion which is usually there may be broken down and warmer maritime air will move in over the ice.

The general circulation aloft over the Antarctic appears to be zonal. Westerly winds prevail at the surface poleward to around $65°$ where they give way to low level easterly winds. These easterly winds extend on to about $75°$. Cyclonic storms develop over the oceans in the Westerlies and move around the Antarctic from west to east. They normally do not penetrate too far inland and they account for much of the precipitation and weather along the coast. Storm activity is largely responsible for the "Kernlose" temperature phenomenon. This is a reversal of seasonal temperature changes. When the temperature first begins to drop after the short summer season, rapid radiational cooling causes a strong on-shore temperature gradient and an influx of warm air results which causes temperatures to rise again for a short time. Then follows a second more permanent temperature drop.

Moisture data are even more difficult to obtain for the polar deserts than temperature data. Data indicate that relative and absolute humidity decrease inland, as does cloud cover. Humidity and cloud cover are also lower in winter as a result of the stronger subsidence and the surface inversion. The cloud decks which appear over the ice caps produce sub-stantial warming of temperatures as they provide a heat source, radiat-ing heat to the ground. Precipitation may be fairly frequent but it is necessarily light. Precipitable moisture in the atmosphere is not very great due to the low temperatures, and probably doesn't exceed 0.4 inches at any one time. Most of the precipitation occurs as snowfall with averages of 30-60 cm (12-24 inches) per year. The water equivalent ranges from 13-25 cm (5-10 inches). The present volume of ice on the Antarctic continent represents an accumulation of some 10,000 years pre-cipitation. Some of the precipitation occurs in the form of ice pellets and some moisture may be deposited directly as condensation from the sub-siding air as it reaches the ice surface. Along the coasts instances of steam fogs drifting inland have occurred when temperatures were as low as $-30°$ C $(-22°$ F).

Optical phenomena are particularly common in the polar regions. The Aurora Borealis and Aurora Australis occur in the respective hemispheres. They are more frequently called the northern lights in the Northern Hemisphere. These are caused by electrical particles from the sun strik-ing the atmosphere of the earth. The aurora are most frequently observed at the time of the equinoxes and peak periods occur concurrently in the two hemispheres. Sun dogs and moon dogs also occur frequently. These are the result of reflection and refraction of light waves from ice crystals in the atmosphere. Mirages are perhaps more common in these areas than in the deserts, due to the strong inversion at low altitudes. White-out is another major hazard to travel near the poles. It is a result of light reflecting back and forth between the ground and clouds in such a way that everything appears white and it is difficult to distinguish between the ground surface and the atmosphere.

Highland Climates

The climatic classification set forth in an earlier section establishes the basic seasonal patterns of the major climates. This classification does not exclude mountain areas. Mountain regions exist within all of the different climatic types. The seasonal regimes of mountain areas and the diurnal patterns are essentially the same as the surrounding land at lower altitude. The major differences that exist between mountain regions and surrounding lowlands are essentially two: temperature decreases rapidly with height, and precipitation amounts change with height.

The change in temperature with height in mountain areas is at a rate roughly a thousand times greater than with latitude. The decrease in temperature results from sheer distance from the primary heat source, the earth's surface. Also significant is the lower ratio of heat retention due to the lower density of the air. The thinner air traps less solar radiation and less earth radiation. The transparency of the air allows for rapid surface heating and cooling but does not greatly affect the overlying air temperature.

Insolation at 4500 meters (15,000 feet) is from 25 to 50 per cent greater than at sea level, due to lower concentrations of water vapor and solids in the atmosphere, and to the longer photoperiod. The amount of shortwave radiation, particularly that in the near visible range of violet and ultra-violet, is increased in even greater than average proportion. The cooler temperatures associated with higher altitudes in the tropics has attracted European settlement to these parts of the tropics. European types of agriculture can be adapted to the higher elevations and also many of the diseases common to the humid tropics are absent or greatly reduced in potency.

Temperature inversions are a common phenomena in mountain basins. The steeply inclined slopes help induce the formation of inversions, by funneling air cooled by radiation downslope with the aid of gravity. The cooler air settling in the valley bottoms may become pronounced, and thick fogs or smogs develop. The inversions are quick to form with the decline of radiation in the late afternoon, but burn away rapidly in the morning with the rise of the sun. When winds are strong the inversions are prevented from forming.

The moisture regime in terms of seasonal distribution and storm type is quite similar to the sea level forms. Mountains themselves, if they lie athwart the direction of the prevailing winds, act as lifting mechanisms and tend to increase rates of condensation and subsequent precipitation. Normally precipitation increases with height in mountain areas for a few thousand feet and then begins to decline. The decline in precipitation that occurs beyond a certain point is due to decreasing amounts of water in the atmosphere as temperatures become cooler. At most locations on the earth's surface the mountainous areas in any climatic region receive the most precipitation. Even in desert regions the higher hills and mountains usually receive significantly more precipitation than lower areas.

The micro-climatic condition at a given site in mountains depends to a large degree on slope orientation. Orientation with respect to the sun is significant in determining local heating characteristics. A slope facing the sun will have much higher surface temperatures and resultant air temperatures than one facing away from the sun, due to the greater intensity of solar radiation.

While orientation to the sun is a primary factor in controlling local temperature conditions within mountains, orientation of slopes with respect to prevailing winds is just as significant in terms of determining local precipitation characteristics. Windward slopes receive up to five times the amount of precipitation that is received on the leeward slopes.

The pattern of winds in mountain areas is as complex as the pattern of radiation and precipitation. The general circulation determines the general direction of airflow over the region. The winds are normally driven aloft by the topography so that winds near the surface, if they are present with any persistency, reflect local wind situations. Several different kinds of winds exist in mountain areas that can be identified as general types.

In mountain areas a diurnal wind system develops along the valley walls and up and down the valley length. During the day the ground surface is intensively heated and the overlying air heated by earth radiation and conduction. The result is the heated air becomes less dense and begins to move up the slope. The air at the bottom of the valley is replaced by subsidence of air in the center of the valley. At night the

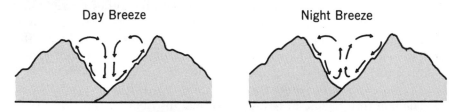

Day Breeze   Night Breeze

FIGURE 8.4 The mountain and valley breeze

ground surface cools faster than the overlying air and the air near the ground slides down the slopes as it cools and becomes more dense, with air converging and rising in valley center, or flowing on down the valley if there is an outlet.

Some winds are largely the result of gravity flow of air. These winds are the flow of cool dry air downward from highlands, either mountain masses or icefields. When air has stagnated over areas of high elevation covered with snow or ice it is chilled considerably. On occasion this air induced by gravity, descends down the topographic slope, bringing low humidity and cold temperatures to the surrounding lowlands. These winds are a characteristic of coastal areas where there are mountains nearby. Such areas are the Riviera of France and Italy, the Adriatic Coast of Yugoslavia, coastal Alaska and Greenland, and coastal Antarctica. In these three areas proper names have been applied to them, indicating the persistence of occurrence. They are respectively the *Mistral, Bora,* and *Willywaw*.

Adiabatically warmed winds also occur in association with mountain areas. These are warm dry winds which descend the leeward sides of mountains. They are warmed by the process of adding heat from condensation as the air rises over the mountains. Normally the air remains aloft but occasionally it descends along the lee flanks of the mountains and, with heat of compression added to the latent heat of condensation, arrives at lower elevations with temperatures which may be considerably above those of the surrounding air. These winds are most noticeable in the winter months when they are capable of raising temperatures up to 22° C (40° F) or more in the space of a few minutes to several hours. Denver, Colorado, has observed a 14° C (25° F) rise in two hours, and Havre, Montana, had a rise of 18° C (33° F) in one hour. As cited previously, Spearfish, South Dakota, experienced a rise in temperature of 27° C (49° F) in two minutes as a result of adiabatic winds. The extreme dryness of these winds is indicated by one such wind in California, in which the air temperature was 14° C (57° F) and the dew point −23° C (−9° F). These winds go under various names in different countries. A partial list is shown below:

Chinook—Northern United States
to Alaska
Santa Ana—California and
Northern Mexico
Foehn—Europe

Latitudinal differences occur in the mountain regimes as well as with lowland regimes. This is expressed in an Equator to poleward decrease in elevation of the snowline and other temperature dependent features. The snowline in mountains near the Equator is found at elevations near 18,000 feet, depending on orientation of the slopes. The snowline drops slowly poleward, until mid-latitudes are reached, and then drops rapidly until it gets near the surface around the Arctic or Antarctic circles.

Atmospheric pressure becomes a significant characteristic in mountain climates. Atmospheric pressure decreases with height at a geometric rate, fast enough so that at around 18,000 feet the pressure is approximately half that of the surface. Density changes at about the same ratio.

Mountain ranges also affect the climates of the adjacent lowland areas by acting as controls for airflow. In mid-latitudes mountain ranges tend to act as barriers to the movement of anticyclones. The cold dense air moving out of polar regions does not move over the mountains very readily, so that there are often sharp contrasts in temperatures from one side of a range to another. This blocking is most pronounced when the anticyclone is moving obliquely to the trend of the mountains.

Mountain ranges also act as precipitation controls on areas on the lee side (rainshadow) of the mountain ranges. Mountain ranges which lie perpendicular to maritime winds reduce the moisture content in the air by causing orographic precipitation on the windward flanks of the mountains. As a result rainshadows develop downwind of the mountains. Some of the deserts of mid-latitudes are largely due to this aspect.

Chapter 9

Weather and
Climate Modification

The present atmosphere is largely a product of life on earth. The large quantity of oxygen which exists in the atmosphere (about 19% by volume) represents the accumulation of oxygen placed there by organisms and not yet used in oxidizing the same organisms. The oxygen is the residual from the formation of the immense fields of coal, petroleum, and natural gas. These hydrocarbons represent the residue of plant and animal matter stored in the environment. Burning of this huge supply of fuels would reduce the oxygen supply in the atmosphere to a small part of what it is today. As man is one of the many forms of life present in the environment he has played a role, as have all species, in the formation of the atmosphere. The atmosphere has been in a continuous process of evolution ever since the first living organisms began to release gases of their own making. Thus, modification of weather and climate has always been a function of the living organisms on earth and hence a function of man's existence. The major variable in the function of man as a modifier of the atmosphere has been one of scale. Until only the last instant or two in the history of life on earth man was present in such small numbers that his influence was largely insignificant compared to the influence of the plant kingdom. The rate at which man has been able to influence the atmosphere is a function of the size of the population and the technological state of the society. Both the size of the world population and the technological state of society have increased at a geometric rate in the last several centuries.

Modification of weather and climate can take place, and has taken place in two different forms based upon man's behavior. Man has deliberately attempted to alter the atmospheric environment to make it more suitable from the standpoint of comfort or economics. This deliberate process probably began when fire was first brought into a cave to heat the damp or chilly air. Man has also inadvertently altered the atmosphere by his very existence as a living species and by his gradual increase in numbers and technology. The increasing amount of heat, particulate matter, and gaseous wastes that are liberated into the atmosphere are exemplary of the latter.

The inadvertent modification of the atmosphere by man has taken place as long as man has been present on earth, and attempts to deliberately alter the weather and climate go back as far as recorded history and undoubtedly beyond that. Primitive methods devised to bring rain or appease the gods of storms are matters of record. In America rain dances were a common part of rituals of some tribes of Indians. In other cases water was ceremoniously sprinkled over young naked girls, and in other tribes small children would be buried in the earth with only their heads showing. Presumably in the latter case the rain gods would take pity on the children, but in the former case the logic is a little more difficult to grasp. It may have been psychologically rewarding if not productive of rain. In many parts of the world it has long been the custom to fire rockets and cannon shells into clouds in an attempt to prevent hail and other characteristics of severe storms. This is still practiced in some areas. These early attempts by professional (?) rainmakers were the beginning of an industry.

Planned Modification of the Atmosphere

Weather modification can be considered as any attempt to alter the natural processes and development of weather systems. This includes attempts to change the rate and amount of precipitation, the suppression of lightning and hail, frost prevention, fog dispersal, and alteration of severe storms such as tornadoes, thunderstorms, and hurricanes. Weather modification has been most successful at the microscale and at the immediate interface of the earth and atmosphere. The primary reasons for the greater success in these situations is the lower amounts of energy needed and the ability to apply the necessary procedures in the right location at the right time.

Microscale Modification

Windbreaks have been widely used to reduce wind velocity at and near the surface. Snow fences are a common sight in the American midwest. The fences constructed parallel to and a slight distance from highways and railways help control drifting snow. They create zones of reduced wind velocity downwind and the snow is precipitated in the lee of the fence rather than along the transportation line. If the windbreak is perpendicular to the wind vector, nearly calm conditions will be produced for a distance of up to five times the height of the break and reduced wind velocities will be attained downwind up to a distance of forty times the height of the break. In addition to controlling blowing snow or sand, wind breaks have been constructed to reduce the pressure of wind on structures. At some large airports wind breaks have been designed to prevent damage to structures by the blast from jet engines. Shelterbelts of shrubs and trees have been widely planted in the plains regions of the world. A shelterbelt will affect the local area in a complex fashion that will involve the air flow pattern over the surface, radiation intensity on the ground, precipitation pattern, evapotranspiration, and alteration of the carbon dioxide exchange.

TABLE 9.1

Approximate Kinetic Energy Contained in Various Atmospheric Circulation and the Time Required to Produce the Equivalent Amount of Electric Energy in the United States

Circulation	Kinetic Energy (Kw-Hrs)	U.S. Production Time
Tornado	10^7	5 minutes
Small Thunderstorm	10^8	50 minutes
Large Thunderstorm	10^9	10 hours
Tropical Cyclone	10^{11}	5 weeks
Extra Tropical Cyclone	10^{12}	50 weeks
General Circulation - Northern Hemisphere	10^{14}	100 years

Source: Miller and Thompson (1970), p. 320.

evaporation suppression

In arid regions loss of water stored for irrigation or other uses is extremely high. Water temperatures are often warm and the air very dry which produces conditions extremely conducive to evaporation. A considerable amount of research has been conducted to find ways of suppressing this evaporation. The most favorable results to date have been obtained by spreading a mono-molecular layer of alcohol over the water surface. Under ideal conditions evaporation has been reduced up to 70 per cent. However, in the real world outside the laboratory such results are not achieved. The best results have been achieved by using cetyl alcohol, a mixture of duodecanol and hexadecanol hydro-carbons. The problem in using the alcohol compounds on lakes and reservoirs results from the low surface tension of the compounds and the natural turbulence that occurs in open water. The alcohol layer breaks at very low wind velocities. To be effective on open water a nearly continuous application of the alcohol is necessary as is application from different points on the surface to compensate for changing wind directions.

frost prevention

In the chapter on moisture in the atmosphere two types of fog and frost were discussed, advective and radiation. Frosts occurring from advection tend to be widespread, heavy and persistent. At the present state of knowledge and operational techniques there is no means for handling frosts of this nature. Radiation fogs of short duration and of a local scale can be reduced and techniques for this purpose are operational at this time. A wide variety of techniques has been developed but they can be classified into several types: (1) the creation of an artificial cloud over the area to reduce the loss of longwave earth radiation-kerosene burners or smudge pots are commonly used; (2) the creation of turbulence to bring in warmer air using large fans or other ventilating devices; (3) the heating of the air by gas or oil burners or spraying water into the air; (4) providing temporary cover for the area constructed from plastic or other material. In the case of frost prevention in areas of high value crops the costs of modification are less than the costs of losing the crop.

fog dispersal

For purposes of fog dispersal two types of fogs are recognized as the methods of treatment for dispersal are different. The two categories are warm fogs and cold fogs with the warm fogs having temperatures above 0°C and cold fogs being those in which at least part of the mass has temperatures below 0° C. Fog dispersal is based on techniques designed to change the size or state of the fog particles and in some cases to change both size and state. Warm fogs can be dissipated by increasing the size of the particles. This has been done by spraying water droplets into the air producing coalescence, and also by creating electrical discharges into the air for the same purpose. Warm fogs have also been dispersed by artificially circulating the air and by artificially heating the air. The latter method has been used to some extent ever since World War II when it was used in Great Britain at some of the air fields.

There has been more success in dissipating cold fogs than warm fogs. The introduction into the cold fog of materials that will change the state of the water particles has provided the greatest success. The seeding of the fog by dry ice has proved effective. The dry ice cools the water droplets to a temperature at which they change to ice crystals. This method is most effective at temperatures below and within a few degrees of freezing. Cooling sufficient to cause freezing has also been accomplished by releasing propane gas through expansion nozzles.

Techniques of both cold and warm fog dispersal are being used at some major airports around the world.

Meso-scale Modification

precipitation augmentation

Interest in artificial rainmaking goes back many years. Professional journal articles have appeared dealing with the topic since 1892 and the federal government has been interested in rainmaking from about the same date. Experiments with seeding of clouds with nucleides began on November 13, 1946 when Vincent J. Schafer, working for General Electric Research Laboratory, seeded a cloud bank in New York.

Most precipitation modification at the present time is based upon the introduction of special nuclei into moisture-rich clouds in an attempt to change the thermal stability of the cloud or the droplet structure of the cloud. Changing the stability or particle structure may increase the amount of precipitation reaching the ground surface. The addition of the nuclei to the cloud initiates or enhances the precipitation process. The various processes and techniques for introducing nuclei into clouds is generally classed as cloud seeding. There are two categories of clouds to be dealt with just as there are two categories of fogs, and the general approach to stimulating precipitation is the same. In warm clouds the process is aimed at increasing droplet size, and this is accomplished by seeding with hygroscopic material. The water absorbent nuclei stimulate the growth of drops to sizes large enough for natural coalescence to take over and produce drops large enough to precipitate. The major problems are in terms of getting the right size of particle in the right condition into the right part of the cloud.

In cold clouds seeding is done with materials that initiate a change of phase, namely to cause the water droplets to fuse into ice crystals. Seeding with ice crystals in supercooled clouds can effect the cloud. It may cool the air and entrained water droplets to the point where spontaneous freezing occurs. The second method of seeding supercooled clouds is with an extremely fine dust containing silver iodide (AgI). The structure of the silver iodide crystal is similar enough to that of ice so that when a particle of the silver iodide compound comes into contact with a water droplet the droplet freezes immediately. In a given cloud there may be both warm and cold sectors so that to obtain maximum increase in precipitation both warm and cold cloud techniques need to be utilized especially at critical times and locations in the cloud system.

Delivery of the various materials is done by either ground to air or air to air techniques. Where dry ice is used to seed supercooled clouds it is almost invariably dropped by aircraft. The size of the ice particles used depends on the vertical height of the cloud. Where silver iodide or related compounds are used they may be either spread from the air or the ground. When ground generators are used natural atmospheric turbulence is used to carry the nucleant to the clouds. Ground generators making use of silver iodide impregnated coke heated to 3000° F have proved successful. At this temperature the smoke particles are extremely fine. Estimates place the desirable release rate at about thirty quadrillion parts per minute. The Soviet Union has been experimenting using artillery and rockets to deliver nucleants to the desired sectors of clouds.

The best results from cloud seeding have been from building cumulous clouds where supercooling existed for several thousand feet, and the cloud was developing in a potentially unstable atmosphere. In such locations precipitation has been increased as much as 10-20 per cent.

In tropical and subtropical areas it has been shown that building cumulous clouds can be forced into violent growth by seeding. The increased rate in growth is due to the addition of the latent heat of evaporation and the heat released in fusion as the water freezes. It was also observed that seeding a single convective cloud had the result of increasing precipitation intensely over a small area at the expense of more general precipitation over a broad area.

Studies of the more abundant cloud forms such as stratus, small cumulus, and strato cumulus indicate that they do not contain enough precipitable water to be of any economic benefit. A cloud 1,000 m thick has a vapor content of about 1 litre per sq meter or a depth of about 1 mm of precipitation. In order to be of any use to agriculture at least 2.5 mm of precipitation on the ground is essential. Too often small amounts of precipitation evaporate before reaching the ground. The curtain of rain hanging below a cloud in such a situation is a *Virga*.

In several precipitation augmentation programs changes in the precipitation pattern have been detected outside the seeding area. While both decreases and increases in precipitation in the peripheral areas has occurred, the general tendency is for an increase. The evidence seems to suggest that seeding affects the dynamics of the local circulation and perhaps upstream as well as downstream from the area of seeding.

Generalized conclusions regarding the results of cloud seeding are difficult. It is apparent that the most common cloud forms will not readily yield water to seeding nor do they contain enough water to make seeding worth the effort. It is even possible to reduce precipitation by seeding when cloud temperatures are too low or when conditions of stability are right. The indications at the present time are that seeding under optimum conditions may speed up the precipitation process and increase the amount of precipitation by 10 to 15 per cent in the immediate area of seeding with lesser changes downstream for distances of up to 250 km. The best results with cloud seeding have been in areas where humidity and natural precipitation rates are both high. Coastal mountain areas in mid-latitudes have provided the best results. It is clear that precipitation cannot be produced where natural conditions do not already favor precipitation by having large amounts of moisture available and lifting mechanisms operating.

hail suppression

Hail suppression has a long history. Explosive shells have been fired into clouds, a practice which is still continuing in some areas. Church bells have been rung incessantly. Maria Theresa of Austria in the 18th century forbade the shooting of clouds by cannon.

Modern experiments have been carried on in a wide variety of environments in the United States and many other countries. Many of these experiments have centered around the seeding of the towering cumulus with large numbers of hygroscopic nuclei to induce precipitation before freezing begins or to inhibit growth of the ice particles.

Micro-scale Atmospheric Modification

A large number of schemes has been suggested at one time or another for altering regional climates and in some cases world climate.

Several suggested projects involve melting the arctic ice pack. One scheme recommended building a dam across the Siberian Strait and pumping the cold water out of the Arctic. The cold water would be replaced by warm water from the gulf stream in the Atlantic. A more recent proposal would melt the ice by spreading soot or coal dust over the ice. If the ice pàck were to melt, the world heat balance would be greatly altered and in all likelihood the general circulation itself would be altered and thus produce a change in world climates. In the Soviet Union fine carbon particles have been spread over snow cover to speed up spring melting. The result has been earlier melt, greater quantities of melt, and deeper thawing of the permafrost.

Two projects have been outlined for Northern Africa. One of the plans recommends pumping Congo River water into the central Sahara where it would form a lake with an area of two million km². The resultant lake would provide moisture for precipitation over the surrounding regions; however, this would probably not be the case since anticyclonic circulation and stability is characteristic of the atmosphere in this area. A second plan would take water from the Congo for irrigation in the Sudan. Not only would additional land be brought into production but evaporation from the irrigation project would increase natural precipitation. Closely allied with this plan for increasing rainfall in the savannas is a plan which would increase evaporation in the Gulf of Guinea—if the water vapor content of the air flowing over central Africa could be increased, the result should be increased precipitation.

The Future of Weather Modification

The degree of future success of weather modification schemes would appear to be closely associated with the scale of the endeavor. Some forms of micro-scale modifications are obviously operational at the present time and expansion of these forms of weather modification can be expected. Techniques of wind reduction, frost prevention, and fog dispersal are examples. While these forms are at a minor scale the cost-benefit ratio is favorable under present economic conditions. For modification of larger scale weather systems the amounts of energy needed and other costs make them impractical. In the immediate future it appears that weather modification for other than micro-scale conditions will be carried on only on an experimental basis. Even on an experimental basis the weather systems attacked will necessarily be at the small end of meso-scale range due to the energy requirements to offset large storms.

Inadvertent Modification of the Atmosphere

Man has had some success at deliberately modifying the atmospheric environment to accommodate his needs. These deliberate effects on the atmosphere are minuscule in comparison to the changes he has wrought in the atmosphere without really trying. The changes which man has produced in the atmosphere are phenomenal and may have gone to the extent of irreversibly changing the climate of the earth. Some forms of change are temporary and hard to document, others have been accurately determined in amount, if not in effect. The changes produced by man whether attempted deliberately or having formed accidentally are often difficult to prove with certainty. The reason being that even under natural conditions, the weather and climate is subject to marked changes through time and space.

Direct Heating of the Atmosphere

Man releases huge amounts of heat from the combustion of fossil fuels. While the amount of heat is large in the sense of absolute amounts, it is small when compared to the heat placed in the atmosphere by condensation or longwave radiation. Considering the relative importance of heat from combustion it will not have any effect on the mean temperature of the earth as a whole. However, since the release of the heat tends to be concentrated at certain points on the earth's surface there will be thermal effects on a local scale.

The Changing Chemistry of the Atmosphere

There can be no doubt that the atmosphere is changing. The discharge of man's waste products into the atmosphere is taking place at an unbelievable rate. It would seem that man would have considered the atmosphere to be a little more precious than he has especially since he can live without it for only a matter of minutes. He has accepted the atmosphere as an inexhaustible resource to which he had absolute rights to do whatever he wished. The human race has been paying the price for this attitude for some time, albeit a greater price in some areas than others.

Most of the effluents discharged into the atmosphere come from the combustion of fuels. Not all of the products of combustion can be considered pollutants unless they reach excessive proportions. The substance placed in the atmosphere in greatest amount is water vapor. The water vapor added to the atmosphere cannot be considered a pollutant as it is a very small amount compared to water added to the atmosphere by evaporation and transpiration. Carbon dioxide follows water vapor in terms of volume of effluents. Carbon dioxide results from the incomplete combustion of carbon. The burning of fossil fuels markedly affects the balance of carbon dioxide to oxygen in the atmosphere. From the hydrocarbon and oxygen are produced carbon dioxide and water. Thus the burning of fuels adds CO_2 and decreases oxygen. The present ratio of oxygen to carbon dioxide in the atmosphere is about 1,000 to one so that a small decrease in oxygen will result in a huge increase in carbon dioxide. If the known commercial reserves of fossil fuels were burned the carbon dioxide content in the atmosphere would triple. If additional reserves of hydrocarbons are discovered and consumed, there will be a proportionate increase in carbon dioxide in the atmosphere.

The exact effect of an increase in carbon dioxide in the atmosphere has not been ascertained. That the effects will be pronounced in the long run is certain. Around the turn of the century P. C. Chamberlain hypothesized that by increasing the carbon dioxide content of the atmosphere a world wide increase in temperature would be produced. Recent estimates from a variety of researchers place the possible increase in temperature up to a maximum of seven degrees F. During the last hundred years there has been no continuous change in world temperatures. Temperatures tended to increase until 1940, but since that time the trend has shifted slightly downward.

A variety of other materials is also placed into the atmosphere. Included are sulfur dioxide, oxides of nitrogen, ozone, ethylene, olefins, and mercaptans. Sulfur dioxide comes from the burning of coal and oil, and it combines with water and oxygen to form sulfuric acid. It is a major irritant to human beings even in very small concentrations. There is no evidence that it is accumulating in the atmosphere, primarily because it is extremely reactive. Its average lifetime in the atmosphere is probably no more than ten days. When the sulfur compounds are precipitated out by rain or snow, they increase the acidity of the runoff. In some cases near large cities acidity of ponds and lakes is being increased to the point that the local ecosystem is being altered. Nitric oxide and nitrogen dioxide are primarily the product of the automobile engine. Ozone is not present in the atmosphere under natural conditions but is formed by the reaction of oxides of nitrogen and organic compounds found in smog when they are exposed to sunlight. Ethylene, found primarily in automobile exhausts, is poisonous to green plants even in very low concentrations. The olefins are a class of hydrocarbons which react with other organic compounds and nitrogen oxides in the presence of sunlight to produce ozone, aldehydes, and a variety of nitrogen compounds. Included in the aldehydes are formaldehyde and acrolein. The mercaptans are organic substances related to hydrogen sulfide. They are extremely odiferous and are associated with Kraft paper mills.

Particulate matter in the atmosphere includes dust which is a nuisance but not poisonous, soot or unburned carbon particles, photochemical materials manufactured from smog, suspended debris, synthetic hydrocarbons in the form of pesticides and herbicides, and radio-active particles from power plants and nuclear weapons.

The diffusion of atmospheric pollutants depends on a variety of factors including the nature of the source, the density and size of the particulates, and the turbulence of the atmosphere.

Sources of pollutants can be considered as point sources such as a volcanic eruption or a smokestack, line sources such as an interstate highway, and general sources such as scattered human dwellings or large cities. On a global scale, of course, the latter source could be considered a point source. The life of particles in the atmosphere varies inversely with size and density. Generally the more dense and larger the particles the more rapidly they will precipitate. The greater the turbulence in the atmosphere the greater is the distance that the pollutant will be transported. Most pollutants are removed from the air by natural processes. Some form nuclei for raindrops, some aggregate in larger particles and precipitate, and some are chemically changed into products considered as non-pollutants or are changed into products which are removed by precipitation.

Dispersion of pollutants depends to a great extent on turbulence with both vertical and horizontal wind components being a factor. Horizontal air flow is the most important aspect of turbulence in dissipating pollutants as it carries pollutants over greater distances and areas than do vertical currents. Vertical motion under normal wind conditions does aid in substantially reducing the pollution level near the surface, however. When circumstances arise in which vertical circulation is reduced pollution levels reach their highest values. When atmospheric circulation is reduced, which is often associated with temperature inversions, pollutants accumulate near ground levels. Situations producing minimal turbulence are radiation inversions, advective inversions, subsidence inversions, and topographic isolation. Radiation inversions occur in almost all geographic locations on occasion, and advective inversions are common throughout mid-latitudes particularly in the winter season. West coasts of the land masses in mid-latitude are areas where subsidence inversions are frequent. These areas are associated with the subtropical high-pressure systems. Inversions associated with topographic isolation are common in enclosed mountain basins and along rivers which have high valley walls.

The Micro-Climate of Cities

Man has been able to modify the atmospheric environment inadvertently to the greatest extent where he has developed an industrialized society and congregated in large numbers. Thus large industrial cities have developed micro-climates with modified atmospheric temperature, precipitation, humidity, wind velocity, visibility, radiation intensity and chemical content. These changes are brought about by the production of heat by cities, the discharge of wastes, or pollution, and the alteration of the surface from vegetation to concrete or asphalt.

Perhaps the best documented form of modification of the atmosphere by cities is by direct heating. Heat is added to the atmosphere by the combustion of fuels and in large urban areas fuel consumption is rather great. The differences in actual temperatures between urban and rural areas varies with a variety of factors including city size, population density, industrialization, and local topographic and climatic conditions.

For cities of 100,000 or more population the mean annual temperature ranges from 0.8° to 2.0° C higher than the surrounding environment. At certain times during the year, the difference between urban and rural areas is much greater than the annual average. Contrasts are greater in winter than in summer. In winter the artificial heating of the atmosphere from the urban area is considerable and the amount of heat added by combustion may exceed the amount added by solar radiation by a multiple of two or three. Partially for this reason mean annual minimum temperatures of large cities tend to be as much as 4° C higher than the surrounding areas. The heat present in and around a city has been fondly referred to as the urban heat island. The heat island makes a significant difference in the frost free season. A significantly larger number of days have frost in the rural surroundings of large cities than in the cities. Differences of as much as 30% have been measured. The result of having fewer frosts tends to increase the length of the growing season in cities. The difference between the last killing frost in spring and the first killing frost in fall may be as much as three weeks longer in a large city. When conditions are ripe for a very severe frost, the frost will occur in rural and urban areas alike. However, in the spring and fall when temperatures just dip slightly below freezing for short periods of time, the differences in frost frequency is significant.

In summer the primary source for atmospheric heat in both urban and rural areas is solar radiation. Since the difference in solar radiation between the two areas is normally small, mean temperatures differ very little. Likewise in summer there is generally little difference in maximum daytime temperatures. A number of studies have shown slightly higher urban maxima while others have shown that when particulate content is exceptionally high the maximum temperatures in the city may actually be lower. Favoring higher atmospheric temperatures in the city is the high percentage of surface occupied by structures and paving. The structures and pavement absorb more radiation than vegetation, so not only is more energy absorbed but less is used in the cooling process of transpiration. The amount of paving in cities is higher than most people realize. Dettwiller noted that in Paris, France, paving amounted to 26 km² out of a total urban area of 105 km², or about 25 per cent of the total area. Evaporation from the surface along with transpiration from plants are both cooling devices. Evaporation is reduced in urban areas as runoff is speeded up and less water is available for evaporation. And quite contrary to popular belief, on summer afternoons it is occasionally slightly cooler in cities when particulate content is high. The amount of particulate matter over a city may be as much as 10 times as great as that over rural areas. The smog reduces the incoming radiation sufficiently enough to suppress the high afternoon temperatures. Contrasts between urban and rural environs are usually greatest in the evening hours in summer, after sunset and often between 10 p.m. and midnight. The rural areas have ceased direct heating by solar radiation and started to cool by radiating heat away from the surface. In the urban area the large amount of energy absorbed by structures and pavement during the daylight hours is being released to the atmosphere. It is not an unusual phenomenon to walk past the side of a large building in the evening and feel the heat radiating from it. As a result of the release of the stored energy after sunset, urban temperatures have been observed as much as 10° C higher than in surrounding areas. Once again the amount of particulate matter has a pronounced effect on the urban heat island. A layer of particulate matter can act much as a cloud layer does, that is as a secondary source of radiation for the city and also as a reflection of solar radiation.

There are differences in the intensity of the urban heat island between normal working days and weekends and holidays. This points out the significance of industrial plants as heat sources. The amount of direct heating of the atmosphere by combustion of fuels on nonworking days may be as low as half of that of working days. The particulate content of the atmosphere over cities varies in much the same fashion, higher on working days and in winter.

The urban heat island is greatest near the city center but usually displaced downwind from the center when wind velocities are light. The temperature gradient is usually greatest near the outer boundary of the suburbs. Temperature gradients may reach several degrees C per km around large cities. There is a vertical as well as horizontal dimension to the urban heat island. The effects of urban heating go upward several hundred meters depending on wind velocities and cloud cover. When wind velocities reach a threshold level, the urban heat island is prevented from developing as the heat is dissipated rapidly. When conditions are calm, the heating may go up several hundred meters. When this situation develops, the heating tends to break down temperature inversions with the result that large cities have fewer inversions than rural areas. Unquestionably a wide variety of factors effect the intensity of urban heat island and one factor that plays a part in extreme situations is topography. Where topography helps to reduce ventilation, the heat island will be intensified. Extreme cases of rural and urban differences in temperatures of as much as 20° F have been observed in isolated situations. The urban heat island develops best in conditions of poor ventilation and stable conditions. These reach their optimum in topographic basins and associated temperature inversions.

It is difficult to determine for sure how much of the long range temperature increases recorded in urban areas is due to heating by the city itself and how much is due to global warming. Global warming over the last one-hundred years or so has been in excess of 1° C.

Richard J. Kopec (1970) studied the urban heat island in Chapel Hill, North Carolina, and found a heat island similar to that of large cities. In summary he stated:

> . . . (1) the nocturnal heat island is evident not only in association with small cities but more explicitly with small non-industrialized cities having low central business district (CBD); (2) only under critical limits of cloud cover and wind velocity will a heat island develop over such towns; (3) significant local relief tends to exaggerate the heat island effect where CBD's and other causal factors are situated at high elevations, and tend to obscure the effect when located at lower elevations; . . . .

The city also effects winds. Landsberg estimated that mean annual wind speeds were as much as 30 per cent lower than rural. The difference between urban and rural wind velocities depend on the season of the year, the wind velocities, and time of day. Tall structures tend to increase friction and hence reduce the velocity of moderate and strong winds. When the urban heat island is pronounced, wind velocities tend to be higher in the city as the heat increases instability. When wind velocities are low, the city tends to promote turbulence and under fairly calm conditions a weak airflow may develop towards the center of the heat island. This system which could be compared to a land and sea breeze would develop only under low wind velocities and may well be offset by friction.

Humidity and precipitation are also different in the urban environment. Differences of up to five per cent have been reported in relative humidity, with the humidity being lower in the city. The differences tend to be greater on summer evenings when the urban temperature is higher and absolute humidity lower. Cloudiness is likewise affected. Large cities have 5-10 per cent more cloudy days and 25 per cent less clear days. The effects are greater in winter and in the morning hours. Precipitation over cities has been demonstrated to be higher but the estimates of the amount of increase vary. A number of processes have been associated with the increased precipitation of cities. Among these factors are the increase in exotic materials in the atmosphere, some of which serve as hygroscopic nuclei; the increased turbulence caused by the heating from the city and the increased friction of the urban structures. Not only is total precipitation greater but increases have been noted in the number of days with precipitation, number of days with thunderstorms, and hail days. Although precipitation is increased over cities, the amount of snowfall is generally less. This is due to the warmer urban temperatures in winter causing more of the precipitation to fall as rain than would otherwise be the case. Perhaps the most extreme example of precipitation increases by man is the La Porte, Indiana weather anomaly. Records indicate that as industrialization around the south end of Lake Michigan expanded, rainfall increased in La Porte. The most noticeable period in the record is from about 1925 to 1965. In the period from 1951 to 1965 La Porte recorded 31 per cent more precipitation, 38 per cent more thunderstorms, and 24 per cent more hail days than did surrounding stations. The magnitude of the changes is in fact so great that the record itself has been suspect as being erroneous. Additional research will hopefully resolve this question. The increase in precipitation at La Porte has been attributed to the large amount of effluents discharged into the atmosphere by the agglomerate of heavy industry in the Chicago-Gary area.

The discharge of wastes into the atmosphere also affects visibility and radiation intensity at ground levels. The effects of atmospheric pollution on visibility are greatest over mid-latitude cities, in the winter months, and on working days. Average annual total radiation may be reduced by 15-20 per cent in winter and 5 per cent in summer. With a high concentration of smog, ultraviolet radiation can be reduced by as much as 90 per cent. During the 1930's coal dust and ash was so thick in the atmosphere in some European cities that total illumination was reduced 40-50 per cent. The shortwave radiation is affected more than the longwave radiation. The smog reduces incoming radiation but it also retards long wave radiation at night and thus acts as a blanket keeping nighttime temperatures higher than they would otherwise be.

The greater effects of scattering can be observed by the difference in sky color. In urban areas the more complete scattering produces a much lighter or whiter color than in rural areas. Rapid changes in sky color are often noticed following a rainstorm which tends to remove much of the debris present in the atmosphere.

The presence of large numbers of condensation nuclei in the form of smoke tends to increase the frequency of fog. In the early part of the century, as industrialization progressed, the number of days with fog increased rapidly, and then declined during the years of the depression. Then with the onset of World War II, the number of days with fog (smog) increased again perhaps reaching its climax in the lethal smog of London in 1952. The smog lasted approximately 114 hours and reduced visibility to less than 10 meters for more than 48 hours. Subsequent deaths due to respiratory problems were estimated at 4000. London and Los Angeles are well-known for their smogs, and they tend to result from different circumstances. London smogs are due to radiation inversions and in Los Angeles they result from atmospheric subsidence enhanced by topographic isolation. These two cities are in no way unique except in terms of the degree of the problem. Many cities in western Europe and the United States have recorded increases in fog frequency of 100 per cent or more.

The problem of atmospheric pollution over cities is greatest when relatively stable atmospheric conditions exist; low velocity winds, little turbulence, high humidity, and temperature inversions favor development of surface concentrations of effluents. Winter mornings are often conducive to smog formation, and the situation is sometimes worsened by the increased heating produced by the firing of industrial furnaces. The added heat starts vertical circulation and debris above the city is then carried downward with the convection currents. It is perhaps worth noting that in the long run the development of the heat island over a city actually helps reduce the level of pollution at the surface. The heat added to the atmosphere raises the height of the inversion allowing the pollutants to be mixed to a greater depth than would otherwise be the case.

References

BATTAN, LOUIS J., *The Nature of Violent Storms,* Garden City, N.J.: Doubleday & Company, Inc., 1962.

BLAIR, THOMAS A. AND FITE, ROBERT C., *Weather Elements,* Englewood Cliffs, N.J.: Prentice-Hall, 1965.

BUDYKO, M. I., *Evaporation Under Natural Conditions,* Washington: Department of Commerce, Weather Bureau, 1963.

———, *The Heat Balance of the Earth's Surface,* Washington: U.S. Department of Commerce, Weather Bureau, 1958.

DETTWILLER, JACQUES, Deep soil temperature trends and urban effects at Paris, *Journal of Applied Meteorology* 9(1970):178-180.

DUNN, GORDON E., AND MILLER, BONNER I., *Atlantic Hurricanes,* Baton Rouge: Louisiana State University Press, 1964.

EDINGER, JAMES G., *Watching for the Wind,* Garden City, New York: Doubleday and Company, Inc., 1967.

FLETCHER, N. H., *The Physics of Rainclouds,* Cambridge: Cambridge University Press, 1962.

FLOHN, HERMAN, *Climate and Weather,* New York: McGraw-Hill, 1969.

HATHERTON, TREVOR, (ed.), *Antarctica,* New York: Frederick A. Praeger, Inc., 1965.

HIDORE, JOHN J., "An Introduction to the Classification of Climate," *Journal of Geography,* 65(1966):52-57.

HUSCHKE, RALPH E. (ed.), *Glossary of Meteorology,* Boston: American Meteorological Society, 1959.

KAHAN, ARCHIE M., *Weather Modification in the 1970's,* Paper No. 4, I.V.C.H., Joint meeting of the American Astronomical Society and Operations Research Society, June 17-20, 1969, Denver, Colorado.

KAHAN, ARCHIE M., STINSON, J. ROBERT, AND EDDY, RICHARD L., *Progress in Precipitation Modification,* Office of Atmospheric Water Resources, Bureau of Reclamation, Denver, Colorado, 1969.

KENDREW, W. G., *The Climates of the Continents,* Fairlawn, New Jersey: Oxford University Press, 1961.

KIMBLE, GEORGE H. T., *Our American Weather,* New York: McGraw-Hill Book Company, 1955.

KOPEC, RICHARD J., Further Observations of the Urban Heat Island in a Small City, *Bulletin of the American Meteorological Society,* 51: 1970.

KÖPPEN, WLADIMIR, *Grundiss der Klimkunde,* Berlin: Walter de Gruyter & Co., 1931.

LANDSBERG, HELMET E., The Climate of Towns, *Man's Role in Changing the Face of the Earth,* Chicago: University of Chicago Press, 1956, pp. 584-603.

——— Man-made Climatic Changes, *Science,* 170(1970):1265-1274.

LEIGHTON, PHILLIP A., Geographical Aspects of Air Pollution, *Geographical Review* 56(1966):151-174.

LUDLUM, DAVID M., "Extremes of Heat in the United States," *Weatherwise* 16(1963): 108-129.

—— "Extremes of Cold in the United States," *Weatherwise* 16(1963)275-291.

McDONALD, JAMES E., "The Coriolus Effect," *Scientific American* 186(1952):72-78.

MASON, B. J., *Clouds, Rain and Rainmaking,* London: Cambridge University Press, 1962.

NEUBERGER, HANS AND CAHIR, JOHN, *Principles of Climatology,* New York: Holt, Rinehart, and Winston, 1969.

PACK, DONALD H., Meteorology of Air Pollution, *Science* 146(1964):1119-1128.

PETERSON, JAMES T., The Climate of the City, *The Climate of Cities: A Survey of Recent Literature,* National Air Pollution Control Administration Publication No. AP-59, 48 pp., 1969.

PRIESTLY, SIR RAYMOND (ed.), *Antarctic Research,* London: Butterworth's, 1964.

REITER, ELMAR R., *Jet Streams,* Garden City, New York: Doubleday and Company, Inc., 1967.

RIEHL, HERBERT, *Introduction to the Atmosphere,* New York: McGraw-Hill Book Company, 1965.

—— *Tropical Meteorology,* New York: McGraw-Hill Book Company, 1954.

ROBINSON, N., *Solar Radiation,* New York: Elseview Publishing Company, 1966.

RUBIN, M. J., "The Antarctic and the Weather," *Scientific American* 207(1962):84-94.

RUSKIN, ROBERT E., "The Measurement of Humidity in Meteorology," *Weatherwise* 16(1963):55-61.

SATER, JOHN E., *The Arctic Basin,* Washington: The Arctic Institute of North America, 1963.

SEMAT, HENRY, *Fundamentals of Physics,* New York: Holt, Rinehart and Winston, Inc., 1966.

SHEAR, JAMES A., "The Polar Marine Climate," *Annals of the Association of American Geographers* 54(1964):310-317.

SUTCLIFFE, R. O., "Water Balance and the General Circulation of the Atmosphere," *Quarterly Journal of the Royal Meteorological Society,* 82(1956):385-395.

THORNTHWAITE, C. W.ARRIN, "The Climates of North America According to a New Classification," *Geographic Review,* Vol. 21, No. 4, October, 1931.

—— "The Climates of the Earth," *Geographic Review,* 23(1933):433-440.

—— "An Approach Toward a Rational Clossification of Climate," *Geographic Review* 38(1948):55-94.

TREWARTHA, GLEN T., *The Earth's Problem Climates,* Madison, Wisconsin: University of Wisconsin Press, 1961.

TUNNEL, G. A., *World Distribution of Atmospheric Water Vapor Pressure,* Geophysical Memoirs, No. 100, British Meteorological Office, 1958.

U.S. DEPARTMENT OF COMMERCE, ESSA, *World Weather Records, 1951-1960* Vol. 3, "South America, Central America, West Indies, The Caribbean and Bermuda," 1966.

WATTS, I. E. M., *Equatorial Weather,* London: University of London Press, 1955.

WEBB, WILLIS L., *Structure of the Stratosphere and Mesosphere,* New York: Academic Press, 1966.

VISHER, STEPHEN SARGENT, *Climatic Atlas of the United States,* Cambridge: Harvard University Press, 1954.

Index